Rationality, Time, and Self

Olley (F.O.C.H.) Pearson

Rationality, Time, and Self

palgrave
macmillan

Olley (F.O.C.H.) Pearson
Department of Philosophy
Durham University
Durham, UK

ISBN 978-3-319-71972-6 ISBN 978-3-319-71973-3 (eBook)
https://doi.org/10.1007/978-3-319-71973-3

Library of Congress Control Number: 2017964130

Cover illustration: knape/Getty Images

Printed on acid-free paper

This Palgrave Macmillan imprint is published by Springer Nature
The registered company is Springer International Publishing AG
The registered company address is: Gewerbestrasse 11, 6330 Cham, Switzerland

For Barbara van Someren (I am sorry it took so long)

Preface

I have enjoyed the philosophy of time since I began my studies as an undergraduate student. This passion has carried its momentum through my postgraduate and postdoctoral research and into this book. The course of this journey has taken me through a number of topics which are not standardly considered in the philosophy of time, such as the nature of rationality. However, I believe, and I hope this book establishes, that it is important to follow such a broad path.

Discussions of the philosophy of time are often linked with discussions of language. I believe that there is something appropriate in this. Our uses of language can tell us a great deal. In particular, they can help demarcate the beliefs that we have and our awareness of the world. However, I have always felt that the cart over takes the horse once we try to account for our uses of language merely through considerations of meaning insofar as this latter can be separated from, or vary independently of, our connection to the world which language enables. Language is invaluable, but we must not forget that this value derives from the connection with the world that language supports. I believe that a consideration of our rational actions keeps the horse ahead of the cart by ensuring this connection remains central. I hope that even if readers are not convinced by all of the arguments in this book, they can nevertheless see the value of the links that I make between time and self on the one hand and our worldly actions and interactions on the other.

I intend this book to provide new arguments for the tensed theory of time and emergentism about the self. To this extent it is suitable for researchers and advanced students. However, I have also tried to write a book that can be followed—enjoyed—by readers less familiar with the topics. To this end the book introduces all of the key topics that it considers. I have also suggested a range of possible further readings in footnotes to specific arguments and topics. But, more experienced readers can use section headings as a guide to skip already familiar issues.

In this book (in particular Chap. 2) I have used material from my previously published paper 'Appropriate Emotions and the Metaphysics of Time', *Philosophical Studies* (forthcoming). This has improved the book and I am grateful to the journal and its editors and publisher for their support. As I indicated above, this book also grew out of my Durham University doctoral thesis.

Just as my passion for the topic has existed for some time, I have also benefited from the advice, insight, and encouragement of individuals going back a number of years (too many to mention but not too many to remember). I would particularly like to extend my gratitude to those in the philosophy departments of the University of Wales, Lampeter and Durham University, and especially Janie Agnew, Stefano Catelan, David Cockburn, Jonathan Lowe, Donnchadh O'Conaill, R. (Rocky) Rockingham Gill, and Matthew Tugby. I must also express my thanks to Beth and my friends and family for their patience when work took over and their kindness when it did not. Thank you.

Durham, UK Olley (F.O.C.H.) Pearson

Contents

1

Introduction

This book has three interrelated topics: rationality, time, and self. I will reveal their connection through a consideration of indexical language and the beliefs it expresses. In particular, their connection will be brought out through a consideration of the role of our beliefs in our everyday actions.

I take the nature of time to be a topic which is of great importance because our world is temporal. Everything we do and everything we know about is in time or at least, depending upon one's opinion of abstract entities, a great many things we know about are in time. Because of time's pervasive nature we can consider it through considering the nature of our lives as rational creatures who are active and emotional.

Tensed language appears to say something different about the world than tenseless language. We appear to know something different about an event when we know that it is future, than we do when we simply know tenselessly that it occurs at a specific time or after another event. We can know the two tenseless things but still not know if the event has happened yet. This difference appears to concern the nature of time. I believe that these appearances are correct. There is an objective distinction between something being future, it being present, or it being past which is captured by tensed language and in this respect tensed beliefs but which

Olley (F.O.C.H.) Pearson, *Rationality, Time, and Self*,
https://doi.org/10.1007/978-3-319-71973-3_1

is not captured by tenseless language or beliefs. To accept this is to adopt a *tensed* theory of time and my primary aim in this book is to provide an argument to show that this theory is correct, A1.

Not everyone shares my opinion and I refer to those who disagree as *tenseless* theorists.[1] The tensed theory relies on two ideas. First, tensed language plays a different role in our lives than tenseless language. Second, this difference in roles involves a difference in what is described by the language. The former of these ideas allows us to distinguish tensed and tenseless beliefs as the different beliefs one portrays with the different language. The latter idea allows us to speak of these beliefs as involving an awareness of different states of the world, or facts.

It is possible to challenge both of these ideas and tenseless theorists have done so. Thus I have two secondary aims in this book. First, I will defend the idea that different roles are played by tensed and tenseless beliefs and language, A2. Second, I will defend the idea that this involves them capturing something different about the world, A3.

Early in the twentieth century the opinion was popular in philosophy that tensed and tenseless language did not play different roles and that they could in fact be inter-substituted for one another. This view went hand-in-hand with the popularity of standard first-order logic which symbolized the two equally. Since then, however, this view has gone out of popularity and today it is widely recognized that the two are not inter-substitutable, but rather differ in meaning. This is a view accommodated by the invention of tense logic.

There are two broad reasons for this change in opinion. Firstly, it has been recognized that tensed and tenseless aspects of language really do play different roles. There are occasions in which the former can be used and the latter cannot. This difference has been noted in a consideration of our everyday actions and emotions. Knowing that I have to cook supper tonight motivates me to go shopping in a way in which simply knowing anything tenseless does not. For example, knowing an obligation for me to cook falls on the 9th of December 2017 might leave me unmoved even though it is the 9th of December 2017 if I do not know that is *now* the date. Similarly, I might be anxious because I know I am going to cook tonight and not because of any tenseless knowledge I have, such that I am going to cook on the 9th of December 2017.

Secondly, advances have been made in the philosophy of language that provide powerful theories of meaning according to which tensed and tenseless elements of language mean different things. There has been a growth in the popularity of direct reference theories of indexical language according to which indexical language, including tensed language, has an element of meaning akin to the rules of use for that language. This element of meaning distinguishes tensed and tenseless language, because whilst one might use both 'now' and '11:15' to pick out the same time 'now' can also be used to pick out other times in a way in which '11:15' cannot.

To this extent A2 is generally accepted. No doubt these two motivations for accepting it are related because two sentences which cannot be used for the same linguistic purposes amongst speakers who understand both clearly differ in meaning.

In denying A3 tenseless theories of language and truth have been offered according to which a difference in meaning can go hand-in-hand with an identity of truth-makers. That is, it has been argued that whilst tensed and tenseless aspects of language differ in meaning the truth-makers of both can coincide. Moreover, we can express the truth-makers of a use of tensed language with tenseless language and hence we can suppose that the former does not capture facts which the latter does not because the truth-makers can coincide. For example, it is suggested that a use of 'it is raining now', said at 11:20 on the 9th of December 2017, will be true if and only if rain occurs [tenseless] at 11:20 on the 9th of December 2017. This latter fact can be captured with tenseless language and therefore tensed and tenseless uses of language can share truth-makers and be taken to capture the same facts.

This allows the tenseless theory to accept A2 and to explain the different roles that tensed and tenseless aspects of language have in terms of their differing in meaning. This can be done without needing to recognize that the two capture different facts because the difference in meaning does not entail a difference in truth-makers. A3 is denied.

However, a fuller consideration of the different roles tensed and tenseless beliefs play in actions and emotions reveals the importance of rationality (or perhaps appropriateness in the latter case). It is not the case that people cannot be motivated by tenseless beliefs in everyday actions. But, very often they would be acting irrationally if they were to be. I may be motivated to shop because I have the tenseless knowledge that I am obliged to cook on the 9th of December 2017. But, it would be irrational of me to be motivated in this way if I do not know what the current date is (and hence have a tensed belief too). If it is now the 10th of December 2017 or the 10th of June 2017 shopping is pointless.

In a related way, my anxiety that I have to cook today is only appropriate if I have to cook today. Whereas, it is not clear it can be appropriate at all to have a tenseless anxiety because we ought only to be anxious about the future and a tenseless anxiety will be one which is seemingly indifferent to this. If it is the 10th of December 2017 today, then I oughtn't to be anxious about having an obligation to cook on the 9th of December 2017.

Since the middle of the last century it has been commonly believed that actions are caused by belief/desire complexes which provide the reasons for those actions. I refer to this view as *psychologism* about reasons. Psychologism easily accommodates a tenseless acceptance of A2 and denial of A3 because it focuses on the nature of the beliefs themselves rather than what they are an awareness of in its account of an action and its rationality.

In more recent times, however, psychologism has come under increasing attack. There is now a trend of recognizing that reasons for action are facts and hence that the role of a belief in a rational action is determined by its object—the facts it is an awareness of. This *non-psychologism* does not sit well with an acceptance of A2 and a denial of A3. Rather, it implies that A2 follows from A3. It is because tensed and tenseless beliefs capture different facts that they can play different roles in rational action. This is an issue apparently unnoticed by tenseless theorists to date.

My method in this book will thus be to argue that *rational* action differentiates between tensed and tenseless beliefs. In this way I defend A2 in a way that goes beyond its already popular acceptance. I will defend a form of non-psychologism about reasons for action, and will use this to argue that the difference in the roles of tensed and tenseless beliefs stems from a difference in the facts they capture. In this way my defence of A3 will be in tune with contemporary achievements in the philosophies of rationality and action, and will thereby bring these into concert with the philosophy of time in a way which is currently lacking to the detriment to each of these fields. A2 and A3 lead to A1 given our actions are by and large rational. The tensed theory of time is thus established through a consideration of our nature as rational animals.

In the last fifty years it has become popular in the philosophy of emotions to think that emotions are not mere feelings and that they also involve cognitive aspects. They are related to beliefs. It is also increasingly popular to speak of the appropriateness of emotions. To the extent that this has come into contact with the philosophy of time it has again been argued that it coincides with a tenseless acceptance of A2 and denial of A3. A form of psychologism is adopted according to which the beliefs involved in emotions can affect their appropriateness independently of the objects of those beliefs.

However, these moves have been made without a consideration of the recent advances made in the philosophies of rationality and action. Inspired by these I will defend a non-psychologism according to which emotions are responsive to facts which are the objects of their cognitive components. These facts are the reasons the emotions are had for and they thereby determine the appropriateness of the emotions. This entails that the role of a belief in an emotion is determined by its object.

Thus I will run a second defence of A3, A2 and A1. I will emphasize that tensed and tenseless beliefs play different roles in justifying emotions, A2. I will support A3 by adopting a non-psychologism in regard to reasons for emotions and in doing so I will advance the philosophy of emotions by bringing it into dialogue with the philosophies of rationality and action. I will defend A1 on the basis that our emotions often appear appropriate and if they are not, this suggests, unacceptably, we as emotional creatures are often wrong about the facts.

My arguments concerning time have relatives concerning the self. It appears that we can make different uses of first-personal and non-first-personal language, B2. One might hence wonder if this is because the two capture different facts, B3. Emergentism is a view of the self according to which it is something over and above the physical body. Plausibly the physical body can be fully described by a science which uses non-first-personal language. Therefore, if B2 and B3 are correct, we have grounds for adopting emergentism about the self, B1.

I will defend B2 whilst defending A2. I must know that *I* am due to cook supper and not merely that Olley or the author of this book is if I am going to act rationally in gathering ingredients. This is because I may not know that I am Olley or the author of this book.

Non-psychologism shows that if two beliefs differ in their rational import this is because they capture different facts. B2 therefore leads us to B3 and B1 just as A2 leads us to A3 and A1. As with the case of time these links have been neglected. Perhaps this is in part because one can accept B2 whilst failing to recognize the unique rational import of first-personal beliefs and in part because until recently psychologism has, erroneously, been the dominate view of reasons.

Thus this book aims to increase our understanding of the nature of time and of ourselves by considering rationality. The three topics which are the focus of this book will hence be shown to be intimately and importantly connected.

Establishing the tensed theory of time and emergentism about the self in this way will rely very little on theories of meaning (where these are something more than just descriptions of patterns of linguistic behaviour). However, it will have great implications for these because it implies that tensed and tenseless elements of language differ in meaning in such a way that they capture different facts. It also implies that first-personal and non-first-personal elements of langugae differ in meaning in such a way that they capture different facts.

1.1 Chapter Outlines

In Chap. 2 I introduce some terminology and clarify the nature of the debate between the tensed and tenseless theories of time. I examine an argument given by Prior which shows that tensed and tenseless language plays different roles. Prior's argument concerns emotions and I argue that considerations of it that have been offered by tenseless theorists fail to do justice to intuitions regarding the appropriateness of emotions. My conclusion is that tensed and tenseless aspects of language play different roles and that in order to get clear about the implications this has for the philosophy of time we must get clear about the appropriateness of emotions.

In Chap. 3 I introduce a little more terminology and examine an argument given by Perry that shows that tensed and first-personal language plays different roles to tenseless or non-first-personal language. Perry's argument concerns actions and I argue that considerations of it that have been given by tenseless theorists fail to do justice to intuitions regarding the rationality of these actions. My conclusion is that tensed and first-personal elements of language play unique roles and that in order to evaluate the implications this has for the philosophy of time we must get clear about the nature of rational actions.

In Chaps. 4 to 6 I take up part of the challenge raised by the conclusion of Chap. 3 and defend a view of the nature of rational action. In Chap. 4 I argue that the reasons that we act for and that justify our actions are facts and not belief/desire complexes or representational entities.

In Chap. 5 I defend a view according to which acting for a reason, at least in a good case, involves: having a belief that is an awareness of that reason; adopting a goal of performing an action which is justified by the reason one is aware of; choosing to perform that action; and performing that action. Further, explaining actions in this way does not make any assumptions about how the actions are caused.

In Chap. 6 I argue that the rationality of an action derives entirely from reasons. It is not a matter of fulfilling normative requirements of other sorts. In particular, it is not a matter of embodying certain patterns of behaviour, such as desiring *x* if one desires *y* and believes *x* is the means to *y*. If patterns such as this are patterns rational behaviour fits, then this is not because these patterns have their own normative import. It is simply because they are patterns that are legislated by reasons.

Chapter 7 brings together much of the preceding discussion. I argue that rationality demands tensed beliefs in place of any tenseless ones and this can only be because the tensed beliefs capture reasons—that is, facts—that the tenseless beliefs do not. This argument also establishes that first-personal beliefs capture facts that no non-first-personal beliefs do. In addition I clarify the notion of fact which I use throughout this work.

In Chap. 8 I show that the argument of Chap. 7 has implications for other elements of language. For example, two sentences which differ only in respect of the names they contain may capture different facts about the world even if those names are co-referring. The same can be said of co-referring natural kind terms and co-denoting definite descriptions. These implications are already independently defended by descriptive theories of meaning according to which reference is determined by descriptive content. They are also supposed more generally by anyone who attempts to account for differences in cognitive significance in terms of associated information, even if this information is not taken to enter the truth-maker of the sentence. These implications are therefore not worrying and in fact have independent support. These fine-grained distinctions in beliefs couple naturally with fine-grained distinctions between actions, though I argue that they only require fine-grained distinctions between desires.

Chapter 9 defends a cognitive view of emotions according to which they represent the world to be a certain way and they are had for reasons. This enables me to apply the lessons from rational action to emotions and it is shown that the reasons emotions are had for are facts which the emotions represent. These reasons can ground our everyday talk of the appropriateness of emotions in a way which other popular approaches to the normativity of emotions cannot. This appropriateness of emotions is not a matter of their pragmatic value or the appropriateness of their cognitive component. Furthermore, when combined with the fact that the tense of an emotion affects its appropriateness this shows that tensed emotions must capture facts not captured by tenseless ones.

The conclusion of these arguments is that tensed beliefs capture facts that tenseless ones do not and that one's first-personal beliefs capture facts that cannot be captured by non-first-personal beliefs or anyone else's beliefs. The former of these conclusions has been attacked on the basis of McTaggart's paradox and in Chap. 10 I show that this attack is mistaken. The second conclusion appears to conflict with Wittgenstein's argument against a private language, though I show that there is no real conflict here or at least to the extent that there is the argument against my conclusion is mistaken. It is also shown that the unique role of first-personal beliefs supports an emergentist theory of the self.

Finally, in Chap. 11 it is made clear why the conclusions which I have drawn differ from those commonly reached in considerations of our need for indexical beliefs. It is shown that my discussion focusses on the rational import of indexical beliefs in a way other discussions do not. This focus explains the difference in conclusions drawn.

Notes

1. I will define the tensed and tenseless views in more detail in Chap. 2.

Part I

Indexicals and Rationality

2

Tense & Emotions

McTaggart (1908, 1927, bk.II ch. 33) highlights that events can be distinguished both as *earlier* or *later* than[1] one another and as *past*, *present* or *future*. For example, my lunch is both later than my breakfast and future. These two ways of distinguishing events give rise to two ways of ordering these events or two series. For example, last night's supper is earlier than this morning's breakfast, which is earlier than my writing this, which is earlier than today's lunch, etcetera. Similarly, last night's supper is past, this morning's breakfast is less past, my writing this is present, today's lunch is future, etcetera. McTaggart called the characteristics that ground this distinction the B- and A-characteristics respectively, and the two series the B- and A-series. A description of the B-series will always be true if it is ever true. Whereas, a description of the A-series may be true at one time and false at another. This evening it will no longer be true to say "today's lunch is future".

This distinction has prompted extensive debate in the philosophy of time concerning whether or not the A-series is real. The focus of this debate does not concern the reality of the linguistic distinctions between the past, present, and future. All parties agree that some entities[2] are described as past whilst others are described as present or future. The

Olley (F.O.C.H.) Pearson, *Rationality, Time, and Self*,
https://doi.org/10.1007/978-3-319-71973-3_2

focus is rather on the metaphysical significance of this distinction. The concern is whether there is an actual distinction between some entity being future and that entity being present or past which goes beyond the way that entity is described. This debate shall be one focus of this book. I will argue that there is more to something's being future, past, or present than its being described or recognized as such. In this chapter I will clarify the debate and some of the moves that have been made in it. I will consider the validity of some of the arguments involved and will show that in order to make progress it is important to say more about the role of tensed language and beliefs in rational action and emotion.

2.1 Tensed Language

It is important at this stage to get clear about language and tense. Much talk of tense concerns the nature of language, for example, the differences between the sentences 'Sue will run a race' and 'Sue ran a race'. These differences enable us to speak of tensed language or tensed utterances, sentences, propositions, beliefs etcetera. Such tensed language is, I take it, genuine and familiar.

Not all language is tensed. For example, the sentences 'two plus two is equal to four' and 'patience is a virtue' are not. The 'is' in these sentences is a tenseless one. With this tenseless 'is' we can also construct sentences concerning time which are tenseless such as 'sunrise is before sunset' or 'the date of Angus's birth is the 10th of July 1981'.[3] In this way we can see a number of temporal expressions as tenseless: they do not entail that the sentences that they occur in are tensed. I will often indicate the tenseless 'is' by placing it in square brackets[4] and unless otherwise stated I will also take this to indicate that the sentence as a whole should be understood as tenseless.

One way to get a grasp on this distinction is to suppose that a tensed sentence is one that says, or says when used, that an entity is past, that an entity is present, or that an entity is future (similarly for utterances, propositions, statements etc.). I take 'the 2017 Tour de France is present', 'the 2017 Tour de France is now', and 'it's the 2017 Tour de France' to equally say that the Tour is present. That the Tour is present, as opposed to past or future, is given by the meaning of each of these sentences even if they do not all use the term 'present'. When someone utters a tenseless sentence such as 'two plus two [is] equal to four' one may be able to infer that an entity is present. If I am the audience of such an utterance I might be able to know that the utterance itself is present. However, although the utterance to this extent implies or even informs one that an entity is present it does not *say* that an entity is present.[5]

A different way to consider the matter is to note that the understanding of a tensed utterance requires familiarity with the context of that utterance in a way in which the understanding of a tenseless utterance does not. I can readily grasp what is expressed by "the date of Angus's birth [is] the 10th of July 1981" without knowing when it was uttered. However, I miss something if I do not know when an utterance of 'it is Angus's birthday' is made.[6]

These comments perhaps leave some grey areas and I will not try to clarify them, but they do clearly classify many everyday examples and are thus adequate to my purposes. On either classification it is obvious that the simple sentence 'I'm making a mess' is tensed and the simple 'a mess [is] made by Olley on 12th of October 1994' is not.[7] Tensed language thus captures McTaggart's A-series though tenseless language does not.

2.2 Metaphysical Tense

There is such a thing as tensed language. However, the issue that I am concerned with is whether time or the world itself is tensed (we might call the former linguistic tense and the latter metaphysical tense). In saying this I do not mean that time or the world must fit the classifications of linguistic tense. The point is rather that there is something about the world uniquely captured by tensed language.

One might in this regard speak of properties of pastness, presentness, and futurity[8] such that the event of my being born has the property of pastness, the event of my writing this has the property of presentness, and the event of my retiring has the property of futurity. Alternatively one might say that only present time or present entities exist[9] or that only past and present ones do.[10] In a different manner one might say that the past, present, and future all exist but that the future is open insofar as there is only one past and one present but there are numerous futures. There are numerous events of my retiring, but only one of my writing this and of my being born.[11]

On all of these views saying that something is present says something about the nature of the world. Linguistic tense picks out metaphysical tense. An implication of this is that an ideal[12] description of the world that utilizes only tenseless language will necessarily be incomplete.[13] It will leave an aspect of the world undescribed.[14] More specifically, an ideal description of the world that utilizes both tensed and tenseless language will more fully describe the world than one that does not use tensed language. I call any theory that has this implication a *tensed* theory and any theory that denies it a *tenseless* theory.[15] In the course of this work I will defend a tensed theory by arguing that there must be aspects of the world which are captured by tensed language that are not captured by tenseless language.[16] (I will generally allow the context to make clear whether I am discussing linguistic or metaphysical tense.)

2.3 The Old Tenseless Theories

If all tensed language could be translated by or reduced to tenseless language, then the tenseless theory would be vindicated because this would entail that tensed language could not tell us more about the world than tenseless language. This view of tensed language has been supported by a number of proponents in theories that take two general forms. (These theories have largely been superseded for reasons I will outline in the following section so I will refer to these as the *old* tenseless theories.[17])

One form of the old tenseless theory suggests that a tensed sentence can be translated by a similar tenseless sentence which includes a date. For example, an utterance of 'Sue is currently happy' made at 2pm on the 2nd of February 2010 can be translated by a 'Sue [is] happy at 2pm on the 2nd of February 2010'.[18]

A more general expansion of what we might call the old *date* tenseless theory runs as follows.

Any '*e* is occurring now' (present tense) said at *t*, where *t* gives a time and date, can be translated by an '*e* [is] occurring at *t*' (tenseless).

Any '*e* has occurred' (past tense) said at *t* can be translated by an '*e* occurs before *t*' (tenseless).

Any '*e* will occur' (future tense) said at *t* can be translated by an '*e* occurs after *t*' (tenseless).

The second form the old tenseless theory takes suggests that a tensed sentence can be translated by a tenseless sentence that refers to itself. For example, an utterance of 'Rhys is happy' might be translated by a simultaneous utterance of 'Rhys [is] happy simultaneously with this utterance'.[19] I will refer to this as the old *token-reflexive* tenseless theory. Expanding more generally we get the following.

Any '*e* is occurring now' (present tense) said at *t* can be translated by an '*e* [is] occurring simultaneously with this utterance' (tenseless) said at *t*.

Any '*e* has occurred' (past tense) said at *t* can be translated by an '*e* occurs before this utterance' (tenseless) said at *t*.

Any '*e* will occur' (future tense) said at *t* can be translated by an '*e* occurs after this utterance' (tenseless) said at *t*.

Both of these theories can be expanded to accommodate other more complex cases too. Reichenbach outlined a way of doing this by distinguishing the point of reference and the point of speech.[20] For example, consider an utterance of 'Stefano will have told Sue' made at 2pm on the 2nd of February 2018. The implication of this utterance is that when it is made Sue has not been told but there is a time in the future such that Sue will have been told prior to that time. This future time is the time of reference. Using this devise the old date tenseless theory might offer the following translation: 'Stefano tells Sue after 2pm on the 2nd of February 2018 and before the time of reference'.[21] Alternatively the old token-reflexive tenseless theory will translate it as 'Stefano tells Sue after I utter this and before the time of reference'. (This time of reference will likely be further picked out in the context in a way that might alter this translation. For example, in a context in which we are discussing a meeting we will have the following week by which time Stefano will have told Sue of your plans this meeting will provide the time of reference.)

2.4 Problems for the Old Tenseless Theories

Mill (1884, bk. I ch. II) took many words to be directly referential. They pick out entities but provide no information about them. This view lost favour as a result of criticisms from the likes of Frege (1892/1980) and Russell (1905) who argued that, for example, it could not account for the informative nature of true identity statements like "Dodgson is Carroll". However, in the mid-twentieth century directly referential theories of language saw a resurgence.[22] In this return to direct reference theories of language were advanced that raised trouble for the old tenseless theories because they denied that tensed language could be translated by tenseless language.

In Kaplan's (1989) theory, for example, the indexical 'now' will refer directly to a time. However, the meaning of this expression is not given simply by this referent—its *content* on an occasion of use—but also by the rules of use of the expression—its *character*. For example, if I utter "it is raining now" at 2pm 12th of February 1997, the 'now' will serve to pick out 2pm on the 12th of February 1997 and will not predicate anything of that time. The expression '2pm 12th of February 1997' also refers to that time and to this extent it coincides with the 'now'. However, even if we suppose '2pm 12th of February 1997' to be directly referential it will still only have the same meaning as the 'now' if it also has the same character as that expression. But, it is clear that the two expressions do not have the same character as they do not have the same rules of use. 'Now' should be used to refer to the time at which it is uttered. '2pm 12th of February 1997' should be used to refer to 2pm 12th of February 1997 whenever it is uttered. It follows that "it is raining now" uttered at 2pm 12th of February 1997 cannot be translated by "it is raining at 2pm 12th of February 1997". Therefore, the old date tenseless theory is wrong.

A similar fate befalls the old token-reflexive tenseless theory. An utterance of 'it is raining now' could only be translated by a simultaneous utterance of 'it is raining at the time of this utterance' if the 'now' and the 'the time of this utterance' share both content and character. But it is clear that they do not. The latter expression is a definite description and hence not directly referential (so that perhaps the definite description picks out the time and says of it that it is the time at which the utterance occurs). Further, it is clear the definite description (and even just the indexical 'this') has different rules of use from the 'now' and so it cannot translate it. Therefore, the old token-reflexive tenseless theory is wrong.

Moreover, given the manner in which the old date and token-reflexive theories were shown to be mistaken it is implausible that any similar alternatives could be correct if we accept Kaplan's theory. This is because it is implausible that there can be any tenseless terms that can share a content and character with 'now'. Tensed and tenseless expressions appear to differ in character because they differ in use. This difference in use gives rise to a second line of difficulty for the old tenseless theories. Even with the particular details of Kaplan's theory put to one side phrases which necessarily have different uses within a competent linguistic community clearly have different meanings.

This second line of difficulty is to my mind most forcefully presented in an argument by Prior which I will refer to as the *thank goodness* argument. Prior considers a case in which he says "thank goodness the exams are over". Not knowing the date Prior would not say "thank goodness the date of the conclusion of the exams is the 15th of June, 1954" even though that was when he was speaking. Prior is also not particularly concerned that he makes that utterance (or has that relief/belief) then and he would not say something like "thank goodness the conclusion of the exams is before this utterance (belief/relief)". In short, the emotion expressed by Prior's tensed utterance is not equally expressed by either tenseless alternative.[23]

In this way Prior forcefully makes the point that in certain cases no tenseless utterance can play the role of a tensed one from which it follows that no tenseless utterance can translate that tensed one. The old tenseless theories are wrong.

2.5 The New Tenseless Theories

In reaction to problems such as these the tenseless theory has taken a different tack. The aim of the *new* tenseless theories (as I shall refer to them) is to admit that tensed language cannot be translated by tenseless language but to deny that one gets at facts not got at by the other. Crucial here is the thought that tensed and tenseless language is made true by the same facts.

As with the old tenseless theories the new ones come in two forms. The *date* tenseless theory (as I will refer to it) can be given a general expansion as follows:

> Any '*e* is occurring now' (present tense) said at *t* will be true if and only if *e* occurs at *t*.
>
> Any '*e* has occurred' (past tense) said at *t* will be true if and only if *e* occurs before *t*.
>
> Any '*e* will occur' (future tense) said at *t* will be true if and only if *e* occurs after *t*.[24]

The *token-reflexive* tenseless theory (as I shall refer to it) can be similarly expanded:

> Any utterance, *u*, of '*e* is occurring now' (present tense) will be true if and only if *e* occurs simultaneously with *u*.
>
> Any utterance, *u*, of '*e* has occurred' (past tense) will be true if and only if *e* occurs before *u*.
>
> Any utterance, *u*, of '*e* will occur' (future tense) will be true if and only if *e* occurs after *u*.[25]

As with their old counterparts, these ideas can be expanded to accommodate tensed expressions of different types and higher orders. For example, we might say a token of 'e was future' uttered at *t* is true if and only if there is a time t^*, such that *e* occurs later than t^* and t^* is before *t*.[26]

The new tenseless theorist's idea is that *e* occurring before, after, or at *t*, or before, after, or simultaneous with *u* is something that can be stated in tenseless language. This enables the new tenseless theorist to say that the facts which make a tensed utterance true are simply tenseless facts. Thus, the tenseless theory is correct. Nevertheless, this does not entail that tensed and tenseless utterances are inter-translatable because an utterance expressing the truth-conditions of a different utterance needn't be taken as having the same meaning as that different utterance.[27]

2.6 Tenseless Theories' Responses to Prior's Thank Goodness Argument

The new tenseless theories do avoid the problems raised for the old tenseless theories by the new direct reference theories of language, such as Kaplan's. However, they must say something further to respond to Prior's thank goodness argument. Crucially, they must explain how it is that Prior can be relieved that the exams are over and not relieved they are prior to his feeling of relief or the date of his feeling of relief even though the exams being over just is their being prior to the relief or the date of the relief (according to the new token-reflexive and date theories respectively). Moreover, they must account for why this emotion of relief is appropriate.

Prior asserts that he is relieved the exams are over but not relieved that they finish before his relief nor relieved they finish before the date of his relief. (I will refer to the former as Prior's *tensed* relief and the latter two as his *tenseless* emotions of relief.) It thus appears that the tensed relief is different from either of the tenseless ones. The obvious manner in which to distinguish these emotions of relief is to say that they have different objects. In the tensed relief one is relieved for the tensed fact that the exams are over. Whereas, in the tenseless emotions of relief one is relieved for tenseless facts, namely that the exams finish on a certain date or are simultaneous with a specific feeling of relief. However, this implies that the tensed fact is different from either of the tenseless ones and hence that there are tensed facts.[28] This is something the tenseless theories cannot accept.

Mellor (1981a, 1981b, ch. 3) initially responds to this by denying that the emotions of relief have an object at all. Rather, the expression of relief given in Prior's uttering "thank goodness" is just an expression of pure relief and is quite distinct from the belief he has and that he expresses in saying "that's over". In fact he could just have well said "that's over; thank goodness" as "thank goodness that's over". But, there remains a connection between the belief and the relief because the belief is true if and only if the exams finished before that belief (or the date of it) and the exams being over before the belief (or date of it) causes the feeling of relief. For

Mellor this excuses the tenseless theorist from the need to postulate tensed facts. They are not needed to account for the truth of the tensed belief that the exams are over because that belief can be given tenseless truth-conditions. Tensed facts are also not needed to be the object of the relief because the relief has no object.

But Mellor notes that there is a further element to the story because the relief is justified after the exams but not before the exams. One shouldn't be relieved a discomfort is over when it isn't. We thus have the question of why the relief is only appropriate after the exams. The tensed theorist has an answer to this. For example, it is because that is when it becomes a fact that the exams are over or past. The tenseless theorists cannot say this because they deny such tensed facts. Nevertheless, Mellor insists that one is mistaken to think that this gives the tensed theorist an advantage because the tenseless theorist has an equal explanation. The relief is justified then because relief is justified when it is after a discomfort and this is a tenseless fact on a par with the tensed one referred to by the tensed theorist.

However, MacBeath (1983) argues that Mellor's initial response is inadequate because it is false to think that the relief has no object: Prior is clearly relieved about something. (This is perhaps more clear if we speak in terms of gratitude. Prior is grateful and not simply grateful—what could pure gratitude be—but grateful for something, namely for the exams being over.) Moreover, if we accept that the relief has an object, it is clearly a tensed one. He is relieved that the exams are over and he is not relieved that they finish before a certain date or before a specific emotion of relief.

MacBeath offers a different tenseless solution. He points out that belief is an essential intermediary between the facts one is relieved about and the relief one has as a result. If Prior did not believe the exams were over he would not be relieved that they were. Moreover, the relief is responsive to the belief not the fact believed because one could falsely believe the exams to be over and resultantly be relieved. The tenseless theorist can thus say that Prior's relief has an object, namely the (supposed) fact that the exams are past, and this is (or would be) a tensed fact. However, it is not necessary for this tensed fact to exist in order for Prior to be so relieved. It is sufficient that he believes it to exist. Moreover, this belief can be true and made true by tenseless facts as outlined by the new tenseless theory. This relief that the exams are over is also justified because relief that the exams are over is justified after the exams (a tenseless fact).

Mellor (1983, 1998, ch. 4) subsequently concedes MacBeath's point that emotions often have objects and accepts MacBeath's proposed tenseless solution. This solution has also been adopted quite generally by tenseless theorists.[29]

2.7 Inadequacies of the Tenseless Theories' Response[30]

MacBeath has rightly argued that Prior is relieved about or for something, namely the fact that the exams are over. I will refer to this as the *object* of his relief. It is that aspect of the world—fact—that he is responding to. This notion of object should be distinguished from the notion of *content*. By the content of an emotion I mean its intentional aspect or meaning, or its propositional content on some understandings of that notion. My notion of content thus differs from Kaplan's notion which was mentioned above.[31] This notion of content is important when considering the new-tenseless theory as it allows one to say that two expressions or propositional attitudes can have different contents but the same object. The Mellor/MacBeath position as I understand it is that Prior's relief has a tensed content but a tenseless object. (MacBeath separates the relief and the belief. As I will make clear in Chap. 9 there is no need to make this separation, rather emotions can have a content as beliefs can. However, I will not take sides on this issue here and so will speak of emotions and beliefs. My argument works on either view.)

MacBeath actually allows for two alternatives: either for tensed and tenseless beliefs to differ in content but coincide in object; or for them to differ in content and object, but for the object of a tensed belief not to exist as tenseless facts can make tensed beliefs true. MacBeath's openness on this is reflected in his discussion of the notion of fact where he offers two alternatives: tensed facts can be understood to be merely intentional (thus rendering tense an aspect of the content), or they can be understood to be such that they are not intentional but such that they are merely supposed facts (thus rendering tense an aspect of the object, but an object that does not obtain).[32] (I will reserve the term 'fact' for such states of the world as occur in MacBeath's latter use. By 'fact' I mean something objective and metaphysically real.[33]) The position I refer to as the Mellor/MacBeath position is the one of these alternatives that I take Mellor and many others in the literature to adopt. However, the fact that MacBeath offers two distinct tenseless answers does not appear to be recognized in the literature. As will become clear, the arguments I give below show both alternatives to be inadequate.

The object of Prior's relief plays an important role in the rationality or appropriateness of his emotion. The object of Prior's relief is the reason for Prior's relief. It is that for which he feels relief and it is that to which he will refer if someone asks him why he is relieved. It is Prior's *motivational reason* for his emotion. On this occasion the fact which is the object of his emotion is also that which justifies Prior's relief. If Prior is relieved the exams are over and they are over, then his relief is justified. The fact is also a *justificatory* reason.[34] (Crucially, these roles are not played by the content of his relief or belief and it is this fact that makes the tenseless response inadequate.)

MacBeath at times speaks of the object of Prior's relief as being a merely intentional entity such that the intentional entity *that the exams are over* can exist even though the exams are not over.[35] But, the object of Prior's relief is clearly not an intentional entity. If we told Prior that the exams were not over he would not see there to be any reason for his relief and his relief would disappear even if we told him that the intentional entity *that the exams are over* still exists. (That the intentional entity would continue to exist is even quite plausible if intentional entities are abstract entities, as they appear to be.) Prior's relief is responsive to the facts not to a merely intentional entity.

Similarly, Prior should not be relieved that the exams are over if they are not over whether or not an intentional entity exists. So it is clear that the justificatory reason for his emotion is also a fact. That the object of Prior's emotion (i.e. his motivational reason) and his justificatory reason are both facts is of little surprise. If motivational and justificatory reasons were thought to belong to two distinct categories of entity, then no one would ever be motivated by something that justified him or her.[36]

The object of Prior's relief is thus not an intentional entity. It is a state of the world or fact. This is clear for two reasons. First, Prior's relief tracks the facts. This tracking is mediated by belief, however, so can be unsuccessful when the relevant belief is false. Second, justificatory reasons for emotions are facts and it must be the case that Prior is motivated by the kind of thing that justifies him if his relief is appropriate.

Emotions are responsive to the world and it is in this responsiveness that their justification lies.[37] This is apparent in our everyday talk and interactions quite generally. It is central to our practices of explaining and justifying emotions. The fact that my grandmother has died justifies and motivates my grief. It leads me to cry and others to take that crying to be appropriate and to comfort me. Furthermore, if she had not died, I shouldn't have been grieving her. The fact that my neighbour has chopped down my tree justifies and motivates my anger, and so on.

Nevertheless, MacBeath is right that Prior might mistakenly believe the exams to be over when they are not over and as a result he might become relieved. The belief that they are over or its content thus appears to play a role in the origin of Prior's relief. However, we have seen that neither the belief nor its content is the object of Prior's relief. Prior is not relieved about the fact that he has a belief or about the existence of some abstract intentional entity. Prior is relieved about the end of the exams, a fact that his belief enables him to be aware of. The belief or its content are not the reason for which Prior is relieved and they are not what justifies Prior's relief.[38]

In summary, MacBeath was right to bring Mellor around to the idea that Prior's relief has an object. However, our everyday practices give this object a role that the Mellor/MacBeath position makes little recognition of. In particular, the object of an emotion is what motivates it and therefore it is also what justifies the emotion if that emotion is appropriate. Further, the entities that play the role of objects are facts and hence not the contents of emotions so long as contents are merely intentional entities. This everyday theory of emotions reveals the inadequacy of a tenseless response to Prior's thank goodness argument as I will now make clear. (I will expand upon this theory of emotions in Chap. 9.)

As outlined above, the second of MacBeath's tenseless responses to Prior's case involved taking the object of Prior's relief to be a tensed fact but denying that that fact obtained (or, more accurately, denying the relief had an object but accepting that if it did it would have been a tensed fact).[39] However, this response to Prior is problematic. In particular, denying that any tensed emotions have existent objects entails that one cannot refer to the object of a tensed emotion to account for its justification or appropriateness, contra the account of emotions just outlined.

According to the everyday account of emotions outlined just above an emotion is appropriate if it is motivated by a fact that justifies an emotion of that sort, that is if the object of the emotion is a justificatory reason for that emotion. It follows that an emotion that lacked an object (or whose object was a non-obtaining fact) would not be an appropriate emotion. A tenseless response to Prior's case that took the line that no tensed emotions had objects would hence entail that no tensed emotions were appropriate. However, there clearly are cases of appropriate tensed emotions because Prior's relief is an example of one. Therefore, this tenseless response is no good. Denying that Prior's relief has an existent object implies that Prior can be relieved because of something and that thing not obtain and his relief nevertheless be appropriate. But this is false. If the exams are not over, then Prior oughtn't to be relieved that they are.

Furthermore, if the (would be) object of Prior's relief that the exams are over is a tensed fact, this entails that Prior's belief that the exams are over serves to pick out that tensed fact. The tenseless theorist's insistence that Prior's tensed belief can be true thus absurdly implies that the belief can be true despite the fact which is its object failing to obtain. But this is clearly wrong. If the tensed fact which is the object of Prior's tensed belief does not obtain then the belief is false. The object here is not merely pragmatically implied by the belief as if it arises as a side effect of the belief or emotion. Rather, it is what the belief or emotion is concerned with.

The point of the last paragraph might be put this way: even if we can state tenselessly conditions which must be fulfilled for a tensed belief to be true, as Mellor argues we can, there still remains the question of whether the satisfying of these conditions is what makes the belief true in the sense of actually grounding the truth. It is clear that it might be that a fact necessarily obtains if and only if a belief would be true without it being the case that that fact is the truth-maker of such beliefs. For example, it is arguably necessary that the fact that one plus one is equal to two obtains if and only if a belief that three plus three is equal to six would be true. But the former fact would not make the latter belief true. Alternatively, the fact that an electron has spin obtains if and if a belief that that electron has charge would be true. But, it is not the case that the fact that an electron has spin is the truth-maker of beliefs that that electron has charge. The point is simply that various facts can be necessarily coincident with other facts and therefore with the truth of certain beliefs without being the truth-makers of those beliefs.[40] Once a tensed fact is recognized as the object of the tensed belief the tensed fact appears to ground the truth of the tensed belief whilst the tenseless truth-conditions merely appear to be such that they necessarily obtain if the belief is true.

If we recognize a tensed fact to be the (would be) object of Prior's tensed belief and we say that that tensed fact does not obtain, then we must say that Prior's belief is false. If Prior's tensed belief is false, then arguably so too are most other tensed beliefs because we can rerun Prior's argument for different events. But this is unacceptable because all of our perceptual beliefs are tensed and so we would be thrown into a thorough scepticism.

Thus, there are two reasons why the tenseless theorist cannot account for Prior's argument by referring to a tensed fact as the object of his relief and denying that fact obtains. First, this would entail that no tensed emotions were ever appropriate which is contra to fact as Prior's appropriate relief makes clear. Second, this would imply that all tensed beliefs were false which is contra to fact as Prior's true belief that the exams are over makes clear.

According to the first of MacBeath's tenseless responses to Prior's case, the Mellor/MacBeath response, a tensed emotion has a tensed content but a tenseless object. I will now argue that that this response to Prior fails.

If we take the object of Prior's relief to be a tenseless fact we will be unable to appeal to this object to account for the justification of Prior's relief or to provide the reason for which he formed that relief. Before the exams are over Prior may well know, and I shall suppose that he does know, that they finish before the date he feels relief (perhaps he knows from past experience that he will feel relieved when they are over). Despite this knowledge, Prior does not feel relieved until the exams are over and nor is Prior justified in feeling relieved that they are over until they are. The problems here stem from the fact that many emotions have a temporal orientation. Relief is orientated towards the past. It is appropriate to be relieved that a discomfort *has ended* but it is inappropriate to be relieved that a discomfort *will end*.[41] In a connected manner, it is inappropriate to have a tenseless emotion of relief because in such tenseless relief one will not know whether or not the discomfort has ended. Just as Prior's relief is inappropriate if the exams are not over, it is also inappropriate if he does not know whether they are.

One could have tenseless beliefs that a discomfort ends at a certain time or simultaneously with a certain feeling of relief or belief, and also not know whether that discomfort has ended. Tenseless relief thus lacks a sensitivity to whether a discomfort has ended yet it is precisely a sensitivity to this that makes relief appropriate. (It is important to note that all parties must agree that it is true to say some things are past and others are not, and therefore these comments do not presuppose a tensed theory of time.)

If no tenseless emotion of relief is appropriate, then no tenseless emotion of relief is justified by the fact it has as object. But, if any emotions of relief have tenseless facts as objects then tenseless emotions of relief do. This entails that no tenseless facts justify tenseless emotions of relief and it implies that no tenseless facts justify any emotions of relief because relief is a single category of emotion.[42] Given the object of an emotion is the sole ground of its appropriateness and a person's tensed and tenseless reliefs could be sensitive to the same tenseless facts—that is share objects—the reliefs' appropriateness must coincide. If no tenseless relief is justified, then neither is any tensed relief. It follows that if one supposes that the object of Prior's relief is a tenseless object, then one cannot use that object to account for the appropriateness of his relief, contrary to the view reflected in our everyday approach to emotions.

Therefore, if we adopt a tenseless theory we cannot account for the appropriateness of Prior's relief in terms of the object of his relief—the reason for which he became relieved. If the object is tensed but does not exist, then Prior's emotion is not appropriate because its object does not exist. Alternatively, if the object is tenseless, then Prior's emotion is not appropriate because tenseless facts do not make instances of relief which have them as objects appropriate. But, it is natural to take the object of Prior's emotion to play the role of making his emotion appropriate. He is appropriately relived if and only if the object of his relief obtains.[43] If one is relieved that the exams are over and the exams are over, then one's relief is justified and appropriate, as Prior's is. Conversely, if they are not over, one should not be relieved. The tenseless theory response to Prior's thank goodness argument is thus inadequate because it has done nothing to address this natural and plausible account of emotions with which it is incompatible.

The Mellor/MacBeath theory does offer an alternative account for why it is that Prior's relief is appropriate. As already noted, it refers to the brute fact that relief about an event that is after that event is justified. But, if this is said by the tenseless theorist we do not have an account that is on a par with the account that a tensed theorist can offer (contra Mellor (1981a, 1981b, 50; 1998, 42)). The tenseless and tensed accounts of the justification of Prior's relief both rely on brute facts. In the former case it relies on the fact that relief is justified after the exams. In the latter case it relies on the fact that relief is justified when the exams are past. But, the tensed theorist offers us more besides this fact. She or he also offers us a story of the manner in which this fact has import to the situation. The fact that the exams are past makes the relief that they are past appropriate because this fact is the object of that relief—the reason it was formed for. The Mellor/MacBeath theorist, however, offers no explanation of this form. As a result, they also offer no explanation of why our everyday talk fits the view of emotion outlined above or of why we say that "Prior is, quite righty, relived because the exams are over".

The Mellor/MacBeath tenseless theory cannot say, as surely we would, that "one ought to be relieved because the exams are over". According to the Mellor/MacBeath theory this utterance picks out nothing to justify relief. According to the utterance the ground for the appropriateness of the relief is given by the clause 'the exams are over'. According to the Mellor/MacBeath account this clause picks out the fact that the exams end before that utterance or before a particular time, *t* (where *t* happens to be the time the utterance was made). However, on the Mellor/

MacBeath theory these facts are not what makes a feeling of relief justified. A feeling of relief is justified if that relief is after the end of a discomfort. At best we have some sort of pragmatic implication of a subjunctive conditional such as 'if you were relieved simultaneously with this statement then your relief would be justified' or 'if you were relieved at time *t* then your relief would be justified'. But there appears to be nothing subjunctively conditional about the idea that one ought to be relieved which is expressed by a statement like "one ought to be relieved because the exams are over". Nor does the instruction appear to be merely pragmatically given. Of course, one might understand an utterance like "one ought to be relieved because the exams are over" less literally and hence in a way more charitable to the Mellor/MacBeath account. However, the point is that reasons must be given for giving this non-literal interpretation and until they are the Mellor/MacBeath account is on the back-foot.

Prior's thank goodness argument thus shows that the tenseless theory is inadequate because it disregards a natural and plausible understanding of the justification of emotions.

2.8 Summary

In this chapter we have been introduced to a distinction between the manner in which entities can be ordered as earlier, later, or simultaneous with one another, and as past, future, or present. This distinction reflects an aspect of language, its form and the way it can be used to describe the world. However, the distinction also has metaphysical import capturing something of the nature of the world itself. If the world is tensed, then we cannot describe it fully with tenseless language alone.

One might believe that the world is tenseless because all tensed language can be translated into tenseless language. But such a view would be mistaken. It conflicts with prevalent theories of the meaning of tensed language. Furthermore, tensed and tenseless language clearly differs in meaning because they differ in possible use.

One might nonetheless believe that the world is tenseless because one believes that all tensed language is made true by tenseless facts. However, this view struggles to account for the role that tensed language and beliefs play in our emotional lives. It appears that our emotions are justified by an object that can only be described with tensed language.

The debate between the tensed and tenseless theories thus invites us to consider in more detail the rationality of our emotions. In the course of this book I will take up this invitation by further investigating and vindicating a view according to which an emotion is justified by its object. I will argue in Chap. 9 that not only is the tenseless theory inadequate because it neglects such a view, but further it is false because it is unable to accommodate it.

Notes

1. I shall also use the terms 'before' and 'after' to refer to these B-characteristics.
2. McTaggart took the time series to be an ordering of events. It is events that are past, present, or future, or that are earlier or later than one another. However, I do not wish to make this commitment so I shall speak simply of entities leaving it open what ontological category these belong to.
3. Lowe (1998a, ch. 4) might deny that these tokens really are tenseless and that they instead express disjunctions. For example, roughly, 'sunrise is before sunset' really expresses something like 'sunrise was, is, or will be before sunset'. However, this would not undermine my argument. It would merely follow that what I describe as tensed sentences capture facts that tenseless ones miss insofar as they say which specific tensed fact obtains (whilst a tenseless sentences would merely say it is one of three options). Nevertheless, I am happy to follow popular consensus and assume that these sentences are tenseless.
4. This is distinct from the use made by Lowe (1998a, ch. 4).
5. This description of tensed sentences does not make a commitment to there being an ontological distinction between an entity being present and it's being future or past, etc., any more than a sentence such as 'the 2017 Tour is present' makes such a commitment. This description is committed to the sentence 'the 2017 Tour is present' saying that the 2017 Tour is present, which given that the object-language and meta-language are in this case one and the same is surely correct. It remains

true whatever one takes the ontological commitments of the object-language and hence meta-language to be and to this extent it is an idea independent of such ontological commitments.

6. Cf. Sider (2001, 12).
7. For alternative classifications, one might consider, for example, Smith (1993, 6–7), or Gale (1968, 42). My overall argument could be made to work with either of these.
8. Cf. e.g. McTaggart (1908, 1927) and Smith (1993).
9. Cf. e.g. Prior (1967) and Bourne (2006).
10. Cf. e.g. Broad (1923) and Tooley (1997, 2003).
11. Cf. e.g. McCall (1994).
12. We should not be bogged down by practical difficulties in compiling such a description.
13. McGinn (1983) (care of Le Poidevin (1991, Introduction)) has argued that a tenseless description will be necessarily incomplete, but that it will not miss out anything in the nature of the world. However, the clause that follows this footnote renders McGinn's point irrelevant and my argument proves this latter positive clause.
14. There is something awkward in saying this insofar as Tooley (1997) and McCall (1994) suggest that all tensed utterances have tenseless truth-conditions. However, my concern is to demarcate a set of tensed theories even if there are also some arguably tensed theories outside that set. Further, it might be a fault of Tooley's and McCall's theories that they take this line (cf. Bourne (2006, ch. 1 II)).
15. Cf. e.g. Lowe (1998b) and Smith (1994) for similar characterizations of the tensed theory. Cameron (2015) tries to avoid using a language focused classification but the theories I classify as tensed will plausibly also be tensed by his standards.
16. One advantage of defining tensed theories in this way is that it leaves open what the tensed nature of the world consists in, for example, whether we must say there are tensed properties of events, or that only the present exists, etc.
17. This is a title used by Oaklander and Smith (1994).
18. Cf. Russell (1906, 256–257) and Goodman (1951, 296). Russell and Goodman were not the only people to support old date tenseless theories. Followers also included, in a manner, Frege (1918/1956) and Quine (1960, sect. 36). However, for Frege it is the tensed utterance and the wider context (including the time it was uttered) that translate a dated tenseless sentence in a manner reminiscent to that in which Russell says

a tensed sentence is incomplete and for Quine the tensed and tenseless do not share meaning but rather are paraphrases of one another that can be used for the same purposes.

19. Cf. Smart (1963, 134) and Reichenbach (1948, 284).
20. Reichenbach (1948, 290). He also adds the point of event but it is not necessary for the example discussed here.
21. Cf. Goodman (1951, 292–293).
22. Cf. e.g. Marcus (1961), Kripke (1981), and Kaplan (1989).
23. Cf. Prior (1959, 1962). Prior actually considered 'thank goodness that's over' but I have chosen to speak explicitly about the exams referred to in order to avoid any complexities added by the demonstrative.
24. Cf. Le Poidevin (1998, 29).
25. Cf. Le Poidevin (1998, 29). Examples of the date tenseless theory can be found in Mellor (1998), Le Poidevin (2007), and Smart (1980). Examples of the token-reflexive tenseless theory can be found in Mellor (1981a), Oaklander (1991), and Dyke and Maclaurin (2002).
26. For an account of how such an expansion might be made see Mellor (1998, ch. 3.5).
27. There are difficulties surrounding these theories specifically in regard to how the formulations outlined in this section are to be interpreted. If we take them as giving truth-conditions, then arguably there is no difference between the date and token-reflexive versions because each state that the token utterance must exist, even if this is only placed after the 'if and only if' of the token-reflexive theory. Further, Lowe has pointed out that one could interpret the stated conditions to be tensed, so that they in effect state: a token *u* of '*e* is now' is now true if it is now simultaneous with *e*, it will be true if it will be simultaneous with *e*, and it was true if it was simultaneous with *e*. The implication of which is that it is not clear that we can state these truth-conditions tenselessly. Cf. Lowe (1998a, ch. 4). However, I do not want to dwell on these problems.
28. We can suppose that if we spoke to Prior he would also deny having any other tenselessly expressible emotions of relief. (Note the difference here. Even if different emotions of relief can share objects, this doesn't make them the same relief. Propensity to express them differently is already grounds for thinking they are distinct.)
29. Cf. Dyke and Maclaurin (2002). It is also accepted by some tensed theorists, for example, Bourne (2006, 17–18). Garrett (1988) adopts a different solution which accepts the existence of the tensed fact but suggests that the tenseless theorist can accept the existence of tensed facts. His

ground for making this assertion is that he thinks that the tenseless theorist is only committed to denying the existence of tensed properties and tensed facts needn't entail the existence of tensed properties. But I think this is a very weak interpretation of the tenseless theory and the existence of tensed facts appears to be an acceptance of a tensed theory (numerous tensed theorists deny that there are tensed properties, for example, Prior).

30. This section makes use of my (Pearson forthcoming) which a reader might look to for a slightly expanded version of the argument of this section.
31. This notion of content may not coincide with other uses of the term in the literature. For example, when Kaplan (1989) speaks of content he means propositional content allowing tensed and tenseless expressions to differ in character but not content. However, for Mellor (1998) and Sider (2001, ch. 1) tensed and tenseless expressions will differ in the propositions they express. Content, as I understand it, is such that Kaplan, Mellor, and Sider all take tensed and tenseless expressions to differ in content.
32. Cf. MacBeath (1983). MacBeath (1983, 87) actually says that these are merely rewordings of one another though it is clear that they are more than that. I suspect a lack of clarity surrounding this issue has not helped matters.
33. Cf. Mellor's (1998, 25) strong, truth-maker notion of fact.
34. Cf. Alvarez (2009) for a discussion of the distinction between the roles reasons might play. I will say more about this in Chap. 4.
35. Cf. MacBeath (1983, 86).
36. Cf. Dancy (2000, ch. 5).
37. This at least characterizes an important class of common emotions (for example, grief, relief, anger, jealousy, envy, gratitude, fear, anticipation, and so on), though I do not mean to assert that it will fit every state that might be thought of as an emotion (perhaps, for example, depression is an exception).
38. As I will make clear in subsequent chapters, especially Chaps. 4 and 9, the fact that Prior could mistakenly take the exams to be over and as a result become relieved needn't entail that the object of his relief is intentional. Rather, one ought instead to say that in such a situation in fact there was no object to Prior's relief. He became relieved without reason. Though, this needn't mean that he did not mistakenly believe the exams to be over or even that we cannot explain his relief in terms of this belief. We can explain emotions without referring to their objects or the reasons

for which they are felt, perhaps, for example, in terms of chemical reactions in the brain.

39. As noted above, I do not take this to be a response that is widely adopted in the literature, nevertheless, it is a tenseless response as it denies that there are any tensed facts and it is a response given by MacBeath so it is worth discussing.
40. Cf. Lowe (2009).
41. Even if it is appropriate to be relieved to have learnt that a discomfort will end, for example, be relieved when one's discomfort at the prospect of endless discomfort ends. Hoerl (2013) in fact distinguishes two forms of relief, temporal and counterfactual. If this distinction were accepted my argument might only apply to the former sort but this is the sort had by Prior and therefore it is the sort that is the subject matter of this book.
42. Relief is an emotion that is a particular type of response to discomfort and made appropriate in this responsiveness. It is common to classify emotions by their objects in this way (cf. Deonna and Teroni (2012, 41)).
43. Actually, the matter is more complicated than this because the object must also be something that justifies the emotion, but in the present case this condition is met.

References

Alvarez, M. (2009). How Many Kinds of Reasons? *Philosophical Explorations, 12*, 181–193.

Bourne, C. (2006). *A Future for Presentism*. Oxford: Oxford University Press.

Broad, C. D. (1923). *Scientific Thought*. London: Kegan Paul, Trench, Trubner & Co.

Cameron, R. (2015). *The Moving Spotlight*. Oxford: Oxford University Press.

Dancy, J. (2000). *Practical Reality*. Oxford: Oxford University Press.

Deonna, J. A., & Teroni, F. (2012). *The Emotions: A Philosophical Introduction*. Abingdon: Routledge.

Dyke, H., & Maclaurin, J. (2002). Thank Goodness That's Over: The Evolutionary Story. *Ratio, 15*, 276–292.

Frege, G. (1918/1956). The Thought: A Logical Inquiry (A. M. & M. Quinton, Trans.). *Mind, 65*, 289–311.

Frege, G. (1892/1980). On Sense and Meaning (M. Black, Trans.). In P. Geach & M. Black (Eds.), *Translations from the Philosophical Writings of Gottlob Frege* (pp. 56–78). Oxford: Blackwell.

Gale, R. M. (1968). *The Language of Time*. London: Routledge & Kegan Paul.
Garrett, B. J. (1988). 'Thank Goodness That's Over' Revisited. *The Philosophical Quarterly, 39*, 201–205.
Goodman, N. (1951). *The Structure of Appearance*. Cambridge, MA: Harvard University Press.
Hoerl, C. (2013). Tense and the Psychology of Relief. *Topoi,* (special issue on *Time and Time Experience*, eds. G. Torrengo & R. Ciuni), 1–15.
Kaplan, D. (1989). Demonstratives. In J. Almog, J. Perry, & H. Wettstein (Eds.), *Themes from Kaplan* (pp. 481–563). Oxford: Oxford University Press.
Kripke, S. (1981). *Naming and Necessity*. Oxford: Blackwell.
Le Poidevin, R. (1991). *Change, Cause, and Contradiction: A Defence of the Tenseless Theory of Time*. New York: St. Martin's Press.
Le Poidevin, R. (1998). The Past, Present, and Future of the Debate about Tense. In R. Le Poidevin (Ed.), *Questions of Time and Tense* (pp. 13–42). Oxford: Oxford University Press.
Le Poidevin, R. (2007). *The Images of Time*. Oxford: Oxford University Press.
Lowe, E. J. (1998a). *The Possibility of Metaphysics*. Oxford: Oxford University Press.
Lowe, E. J. (1998b). Tense and Persistence. In R. Le Poidevin (Ed.), *Questions of Time and Tense* (pp. 43–59). Oxford: Oxford University Press.
Lowe, E. J. (2009). An Essentialist Approach to Truth-Making. In E. J. Lowe & A. Rami (Eds.), *Truth and Truth-Making* (pp. 201–216). Stocksfield: Acumen.
MacBeath, M. (1983). Mellor's Emeritus Headache. *Ratio, 25*, 81–88.
Marcus, R. B. (1961). Modalities and Intensional Languages. *Synthese, 13*, 303–322.
McCall, S. (1994). *A Model of the Universe*. Oxford: Oxford University Press.
McGinn, C. (1983). *The Subjective View: Secondary Qualities and Indexical Thoughts*. Oxford: Oxford University Press.
McTaggart, J. M. E. (1908). The Unreality of Time. *Mind, 68*, 457–474.
McTaggart, J. M. E. (1927). *The Nature of Existence* (Vol. II). Cambridge: Cambridge University Press.
Mellor, D. H. (1981a). *Real Time*. Cambridge: Cambridge University Press.
Mellor, D. H. (1981b). Thank Goodness That's Over. *Ratio, 23*, 20–30.
Mellor, D. H. (1983). MacBeath's Soluble Aspirin. *Ratio, 25*, 89–92.
Mellor, D. H. (1998). *Real Time II*. London: Routledge.
Mill, J. S. (1884). *A System of Logic, Ratiocinative and Inductive*. London: Longmans.
Oaklander, N. (1991). A Defence of the New Tenseless Theory of Time. *The Philosophical Quarterly, 41*, 26–38.

Oaklander, N., & Smith, Q. (Eds.). (1994). *The New Theory of Time*. London: Yale University Press.

Pearson, O. (forthcoming). Appropriate Emotions and the Metaphysics of Time. *Philosophical Studies*. https://doi.org/10.1007/s11098-017-0944-z

Prior, A. N. (1959). Thank Goodness That's Over. *Philosophy, 34*, 12–17.

Prior, A. N. (1962). The Formalities of Omniscience. *Philosophy, 37*, 114–129.

Prior, A. N. (1967). *Past, Present and Future*. Oxford: Oxford University Press.

Quine, W. V. O. (1960). *Word and Object*. Cambridge, MA: Massachusetts Institute of Technology Press.

Reichenbach, H. (1948). *Elements of Symbolic Logic*. New York: Macmillan.

Russell, B. (1905). On Denoting. *Mind, 14*, 479–493.

Russell, B. (1906). Review of 'Symbolic Logic and Its Applications' by Hugh MacColl. *Mind, 15*, 255–260.

Sider, T. (2001). *Four-Dimensionalism*. Oxford: Oxford University Press.

Smart, J. J. C. (1963). *Philosophy and Scientific Realism*. London: Routledge & Kegan Paul.

Smart, J. J. C. (1980). Time and Becoming. In P. van Inwagen (Ed.), *Time and Cause* (pp. 3–15). London: Reidel.

Smith, Q. (1993). *Language and Time*. Oxford: Oxford University Press.

Smith, Q. (1994). General Introduction: The Implications of the Tensed and Tenseless Theories of Time. In Oaklander and Smith 1994: 1–14.

Tooley, M. (1997). *Time, Tense, and Causation*. Oxford: Oxford University Press.

Tooley, M. (2003). Basic Tensed Sentences and Their Analysis. In A. Jokic & Q. Smith (Eds.), *Time, Tense, and Reference* (pp. 409–447). London: Massachusetts Institute of Technology Press.

3

Indexicals & Actions

In the last chapter I introduced the metaphysical notion of tense and linked it to linguistic tense. I outlined the tensed and tenseless theories, and presented Prior's (1959) thank goodness argument in support of the tensed theory. Although the tenseless theory has responded to Prior's argument this response is inadequate as it fails to address a natural view of the rationality of emotions. Specifically, it fails to account for the fact that emotions are justified by the reasons that they are felt for.[1]

The force of the thank goodness argument rests on the fact that the objects of some emotions can only be expressed with tensed language. In this chapter I will show more generally that a large number of our actions require their agents to have tensed beliefs: the reasons the agents act for can only be captured in tensed beliefs. This conclusion will follow from a consideration of arguments presented by Perry (1979) which show more broadly that in the majority of actions the reasons an agent acts for can only be captured by beliefs the expression of which necessarily involves an indexical.

I will argue that *indexical* beliefs, as I shall call them, are specifically required for rational actions. One could act irrationally without indexical beliefs. I will examine an account given by Mellor as to why tensed beliefs

Olley (F.O.C.H.) Pearson, *Rationality, Time, and Self*,
https://doi.org/10.1007/978-3-319-71973-3_3

might be required for actions. I will show that, as with Chap. 2, this tenseless account of the need for tensed beliefs is inadequate because it fails to accommodate a natural understanding of the rationality of the actions at hand. These two chapters thus impress the importance of being more clear about the role of tensed beliefs in rational action and emotion (I will take up this task in the rest of the book). I will begin by making clear some issues of terminology.

3.1 Indexicals

Indexicals are words that can change their referent from one use to another.[2] A common example is the word 'I'. When I utter it it refers to me, Olley, but when my friend utters it it refers to him, Ben. Indexical expressions include 'now', 'yesterday', 'soon', 'I', 'this', 'that', 'he' 'she', 'you', 'here', 'there', 'over yonder', and many more besides. I think that many of these can be grouped. 'He', 'you', 'she', and others like them appear to all be ways of referring to people, I shall call these *personal indexicals*. 'I', 'me' etcetera should perhaps also be considered in this group, though I shall treat these separately as *first-personal indexicals*. 'This' and 'that' and others like them are all ways of referring to objects (in a broad sense) and I shall call them *objective indexicals*. (They are also commonly referred to as demonstratives on the grounds that they are often accompanied by a demonstration of their referent. I shall also follow this terminology on occasion, though plausibly other indexicals such as 'there' are also demonstratives.) 'Here' and 'there' are ways of referring to places, *spatial indexicals*. And of course, 'now', 'yesterday' and 'soon' are ways of referring to times, *temporal indexicals*. (These distinctions are meant to ease discussion though they should not be taken to be exhaustive or exclusive.)

All tensed language shares characteristics with indexicals insofar as it will refer to different things on different occasions of use. For example, "Simon is running" said on the 12th of July 2020 will refer to Simon's running on that date, whilst "Simon is running" said on the 12th of September 2005 refers to Simon's running on the 12th of September 2005. Thus, there are grounds for saying that all tensed expressions are indexicals.

However, it has been argued that 'now' is unlike present tense verbs and copulas, or 'present', as can be revealed by placing these expressions within the scope of a temporal operator. Consider three utterances said on the 24th June: U1 "Tomorrow it will still be the case that it is raining", U2 "Tomorrow it will still be the case that rain is present", and U3 "Tomorrow it will still be the case that it is raining now". If 'now', 'present' and the simple present tense were all equivalent we would expect these three utterances to say the same thing though it appears that they may not. Rather, it has been suggested that the former two say that rain will occur on the 25th of June, while the latter says that rain occurs on the 24th and says nothing of whether or not rain occurs on the 25th. On these grounds 'now' is described as a *pure* indexical while we might call 'present' and present tensed copulas and verbs mere *quasi*-indexicals (because they require double indexing in the scope of a temporal operator).[3]

I think that there is something to these observations, however, I do not take this to deny the context sensitivity of these quasi-indexicals. Rather, it delimits it. Thus, in the following I shall refer to all tensed expressions as indexicals and therefore shall use 'temporal-indexical' and 'tensed expression' to pick out many of the same expressions. This will make matters more simple in most cases as the indexical/quasi-indexical distinction often makes little difference to the arguments at hand.[4]

3.2 Perry's Argument for the Essential Indexical

Perry (1979) has presented an example which I will adapt to concern Tom. Tom was once in a supermarket which had a trail of sugar on the floor. As a conscientious member of the public he began to follow this trail in order to warn whoever was making the mess that he or she had a torn sack in his or her trolley. He followed the trail around a tall counter a number of times before suddenly stopping and rearranging the shopping in his own trolley.

Tom believed all along that the shopper with the torn sack was making a mess and that he or she should do something about it. His sudden change in behaviour is explained by his forming the new belief 'I am making a mess'. The central matter at hand, as Perry points out, is how to characterize this new belief. (I will pick out a belief that someone would express by saying, for example, "I am making a mess", by saying that he or she believes 'I am making a mess'. I will also refer to a belief like this as a *first-personal* belief as it is expressed using a first-personal indexical. I will label other beliefs analogously.)[5]

Tom might well have a belief which is similar but which refers to Tom in a different manner and not be motivated to stop and rearrange his shopping. Suppose we think that Tom forms the belief 'Tom is making a mess'. This belief would only explain his behaviour if he also believed that he himself is Tom, that is, if he had the belief 'I am Tom'. This point can be emphasized by noting that Tom's behaviour might also be brought about by the belief 'Simon is making a mess' so long as Tom also believes 'I am Simon'. Similarly, if Tom believes 'the person wearing a purple and black hat is making a mess', this will not explain his behaviour even if the description uniquely applies to him unless he also believes 'I am the person wearing a purple and black hat'. If he didn't have this last belief, for example, he would simply set off looking for the person wearing a purple and black hat, and not stop and rearrange his own shopping.

It is possible that there are mirrors in the supermarket and that Tom catches a glimpse of himself spilling sugar. He might hence form the belief 'he is making a mess', or 'that man is making a mess', or, if this is different, a *de re* belief of that man which he sees that he is making a mess. Nevertheless, these beliefs will also fail to explain Tom's behaviour unless he has a first-personal belief such as 'I am the man in the mirror'.[6] The indexical 'I' or another first-personal indexical is essential to the belief that explains Tom's behaviour.[7]

Perry makes a similar point concerning other types of actions and other indexicals. Consider Mike who intends to go to a departmental meeting on time and knows that it begins at 2pm. Nevertheless, as the meeting starts Mike remains sat at his desk. Suddenly, as he realizes the time, he stands up to leave.

Mike believes all along that the meeting starts at 2pm and all along intends to attend the meeting at 2pm. What explains his finally moving is his forming the new belief that 'it is 2pm now'. If we attribute to Mike a different belief which replaced the 'now' by another non-indexical reference to the time, then we cannot explain his behaviour. For example, if we attribute to Mike the belief 'it is 2pm at 2pm' this would do little to account for his behaviour. Similarly, if we attribute to him the belief 'it is 2pm at the time when Tom realizes he is spilling sugar', this will only explain his behaviour if we also attribute to him the belief 'Tom is realizing he is spilling sugar now'. The indexical 'now' or a similar present tense temporal indexical is essential to the belief that explains Mike's behaviour.

A third type of case arises for Giulia who is lost in the wilderness despite being an authority on the area. Giulia desires to return home, she is near Llyn Teifi, and she knows that her best way home from there is to head west. However, Giulia continues to look around herself confused until she forms the belief 'this is Llyn Teifi' at which point she heads off to the west. No belief which lacked an indexical could explain Giulia's behaviour though a spatial indexical belief such as 'here is Llyn Teifi' would. For example, Giulia might well believe 'Llyn Teifi is Llyn Teifi' or 'the lake nearest to Strata Florida is Llyn Teifi' and yet lack a motivation to head west.

These examples show that beliefs on which agents act in certain circumstances must be expressed with indexical language. I will refer to the argument for this conclusion given by the use of such examples as these as the *argument for the essential indexical.*

3.3 Assumptions of Perry's Argument

In this section I will make clear a number of aspects of the argument for the essential indexical which are presupposed but not necessarily explicit.

The force of Perry's argument concerns the explanation of actions. The fault of Tom's believing 'Tom is making a mess' is that this belief would fail to explain his behaviour. Thus, we can note that Perry must be assuming that Tom's beliefs explain his behaviour.

On one level it seems quite simply false to say that everything that people do must be explained by their beliefs. A clear example of this is their beating their heart and a less clear example is perhaps their shivering in the cold. However, it seems very important that a great many of a person's doings must be explained by that person's beliefs. These are things that the person does quite deliberately and for a particular reason. They are doings that are intentional and are the result of the will. Such doings are often known as *actions*. Actions are doings or behaviour that stand people out as rational animals.[8] We can be brought to understand a person's action by learning what the person's reason for acting was.[9]

The link with beliefs thus stems from the fact that actions are intentional and are doings for which a person has a reason. To know someone's reasons for an action is to know (some of) the person's beliefs and often vice versa. Tom's reason for acting as he did was that he himself was making a mess. Similarly, the belief of his that explained his action was the belief that he himself was making a mess. It was his belief 'I am making a mess'. This belief explains the action as it provides Tom's reason for performing it. The explanation is of a rational form: we take an individual's making a mess to be a good reason for stopping and rearranging his or her shopping; we take Tom's doing to be an action and something for which there are reasons; we take Tom to be acting rationally; thus, we expect this reason to play a role in explaining his action. We expect this reason to be what he acts for.

Giving the reasons that a person performed an action for can provide an explanation of that person's action. We can use reasons to predict and understand a person's actions. This is not the only form of explanation. It would be quite inappropriate—well impossible—to account for a stone's rolling through providing reasons it acted for, that is by providing the reasons that rationalize its doing. Rather, we should likely give some form of causal explanation, such as the stone rolled because it was kicked by Dr. Johnson. This is why, as was noted above, the rationality of humans is often referred to distinguish them from other entities. This is not to deny that reasons can be causes of actions,[10] but rather to deny that all effects are the result of reasons. (There is a use of the word 'reason' according to which Dr Johnson's kicking the stone was the reason it moved. However, this use differs from the one at hand precisely as it plays no role in rationalizing the stone's movement.)

There are two potential ways of reading this role of beliefs in the explanation of actions. One might, as Davidson (1963/2001a) has, take this to mean that the beliefs actually are the reasons. For example, in Tom's case one of his reasons for acting was his belief 'I am making a mess'. Or, one might take it that the reasons are aspects of the world which the agent is aware of through his or her beliefs. For example, Tom's reason for acting was that he himself was making a mess. Tom's believing 'I am making a mess' is simply his awareness of this fact and required because one must be aware of reasons in order to act for those reasons.[11] This distinction does not affect Perry's argument that indexical beliefs are essential, though it is an important distinction that I shall make use of in time. Whether or not one takes beliefs themselves to exhaust the category of reasons one can still take beliefs to be essential to a rational explanation of action.

The cases at hand in the argument for the essential indexical are therefore specifically cases of actions. They are cases in which an agent acts intentionally for a reason which is provided by or through that agent's belief. Furthermore, these are specifically cases of rational actions. Suppose that Tom forms the belief 'pigs are mammals' and for that reason stops to clear up his mess. In such a case Tom will be acting and his belief will be playing an explanatory role. But, Tom will lack a relevant first-personal belief such as 'I am making a mess'. Crucially, however, Tom will be acting irrationally.

Such a case is possible. People can act irrationally doing things for reasons which are in fact no reason to do those things. People can act for bad reasons. In fact, occasional irrationality of this sort is apparently widespread amongst people.[12] Irrational actions therefore do not require indexical beliefs in the way that rational actions do. The argument for the essential indexical should thus be taken to show that indexical beliefs of particular types are required for *rational actions* of particular types.

The necessity for an indexical belief for some rational actions doesn't entail that all beliefs that play a role in bringing about an action must be indexical. For example, if I am asked "what is two plus two?" I will answer "four". This answer will be in part brought about by my non-indexical belief 'two plus two equals four'. This is clearly a reason for my action and partially explanatory of it. I probably would have said something different if I had not had that non-indexical belief.

It will be useful to distinguish the types of cases which demand indexical beliefs. Plausibly there is something about Tom's case which means that in order to act rationally he must have a first-personal belief.

I will refer to all cases with this property as *personal actions* and to the reasons involved as *personal reasons*. Plausibly, though I rest little on this assumption, the distinguishing feature of such cases is that the agent has a reason for themselves in particular to act and not just for anyone to. Again, there is plausibly something specific to Mike's case that means that in order to act rationally he must have a temporal indexical belief. I'll refer to cases with this feature as *timely actions* and to the reasons involved as *timely reasons*. The linking feature of cases of this form appears to be that they all involve reasons to do something at a specific time and not simply at any time or other. Again, similarly, in cases like Giulia's I'll speak of *locational actions* and *locational reasons*. Here perhaps one has a reason to perform an action at a specific location rather than just anywhere.

Thus there are a number of characteristics or presuppositions of the argument for the essential indexical that we can and should be more clear about. The argument for the essential indexical specifically shows that a rational personal action requires an agent to have a first-personal belief, a rational timely action requires its agent to have a tensed belief, and a rational locational action requires its agent to have an indexical belief. (I will often simply speak of actions for ease of presentation and will instead make it clear when the action at issue is an irrational one.)

3.4 The Sufficiency of First-Personal & Temporal Indexicals

In this section I will consider the extent to which first-personal and tensed beliefs are essential to the types of actions at issue. We saw in the last chapter that the old tenseless theory supposed that tensed language could be translated by tenseless language. We also saw that this view was mistaken but with it in mind we might ask ourselves if the tensed beliefs at issue in the argument for the essential indexical could be replaced by tenseless ones. The argument for the essential indexical clearly shows that the old date tenseless theory is wrong. Mike can know '2pm is 2pm' and related facts, and not be motivated to leave his office. But one might wonder, as perhaps Russell (1940/1995) would have, if Mike could make do with a token-reflexive belief such as '2pm is the time of this belief'. Analogously, one might wonder if Tom could make do with a token-reflexive belief such as 'the believer of this belief is making a mess'.

That token-reflexive beliefs will not suffice for Tom's and Mike's cases can be shown by briefly considering the nature of beliefs. Arguably a belief has a physical manifestation which is perhaps a state of the brain. It would be possible therefore to open up Mike's head and to get a series of images of his brain states and thereby his beliefs. What is crucial here is not that the beliefs are brain states, but simply that they are physically manifested and so can be known third-personally as well as first-personally. But, brain states are our most plausible candidates so I will work with these.

Let us suppose that we in fact implant a micro-camera in Mike's head to take images of his beliefs/brain states in this way and that we display these on a screen in a lab. If this was done we could bring Mike to believe of the belief that he sees on the screen that he has that belief at 2pm. If the images on the screen were live, the belief Mike had could be the token-reflexive and true belief '2pm is the time of this belief'. However, if Mike does not know that the images are live, then no belief like this could replace the essential tensed belief. No belief like this could motivate Mike to leave the laboratory in order to attend his 2pm meeting. (Perhaps Mike thinks the image must be of a past belief and hence that it is too late to bother leaving, or perhaps he thinks it is somehow an image of a future would-be brain state and hence that he still has plenty of time before he must leave.)

Tom might have a similar portable device and know that he is not the only one with such a device so that some of the images of brain states/beliefs that he sees are images of other people's beliefs.[13] Tom might then form a belief that the believer of the belief that he sees on the screen is making a mess. If the belief was Tom's, then Tom would have a true token-reflexive belief 'the believer of this belief is making a mess'. However, Tom will remain unmotivated to stop if he does not know the belief is his. Similar cases could be constructed if one instead spoke of token-reflexive utterances or demonstrative reference to other things besides the beliefs themselves because plausibly these will all have physical manifestations.

So no token-reflexive or demonstrative belief could replace the tensed belief in Mike's case or the first-personal one in Tom's case, and so too for other timely and personal rational actions.

One is perhaps led erroneously to the idea that token-reflexive beliefs could replace the tensed and first-personal ones in these cases because we are often aware that our beliefs are our own beliefs and are present beliefs. What these examples in part show is the extent to which this awareness can be considered. If we were to describe Mike's awareness in a normal situation that his belief '2pm is the time of this belief' is present, it is clear that we ought not to describe it as the belief '2pm is the time of belief x' (where 'belief x' refers to Mike's '2pm is the time of this belief'). Because that would mean this awareness didn't provide the form of presence[14] at issue as the argument for the essential indexical shows. It would also not suffice to describe it as the belief 'the time of this belief is the time of that belief' (where 'that belief' refers to Mike's '2pm is the time of this belief' and similarly below) because this also wouldn't provide us with sufficient form of presence, as I have just argued, and could at best venture us on a regress. The best characterization of this awareness would therefore clearly be as a tensed belief 'that belief is current'. I'm not sure that speaking of this awareness of the presence of our beliefs in this manner is generally illuminating but it is clear that it cannot help one who wishes to argue that token-reflexive beliefs could replace Mike's tensed belief in his rational action. The point is, if we are to refer to this normal awareness we have of the presence of our experience, then we ought to characterize it and if we do, then we ought to characterize it as tensed and hence the essential tensed belief reappears. An analogous argument can be presented for Tom's case. If we are to characterize the normal awareness he has that his belief is his own we could only plausibly do this through a first-personal belief. Token-reflexivity therefore cannot remove the need for tensed or first-personal beliefs.

One might have a further worry in the tense case, specifically: it appears that demonstrative reference involves perception[15]; perception can only be of what is present; and therefore, one can only demonstratively refer to something present; and therefore, one can know of what they demonstratively refer to that it is simultaneous with that reference.[16] However, this worry is unfounded because one can refer demonstratively but in a mediated fashion just as in my example of pointing to an image of something and thereby referring to that thing not the image.[17] We do this all the time, for example, pointing to a television and saying "she is a great actor" and thereby talking of the actor not the image or television. Moreover, perception gives rise to present tensed beliefs.[18] Thus, if we accept that in believing '2pm is the time of this belief' Mike is aware the belief is simultaneous with his perception of it, we must also accept that Mike has the present tensed belief 'that belief is present'. The question thus arises whether Mike's motivation to act relies upon this tensed belief and the problems already outlined imply that it does. Knowing 2pm is simultaneous with one's belief and that this belief is simultaneous with a perception of it provides no motivational force. Rather, one must actually be perceiving and having the tensed belief to be motivated. In short, to the extent that demonstrative reference entails perception and therefore might enable Mike to act it also gives rise to tensed beliefs and these are essential to the action and thus the essential indexical reappears.[19]

Tensed beliefs are required for rational timely actions and first-personal beliefs are required for personal actions. Tensed beliefs and first-personal beliefs are therefore irreplaceable in their roles. I will now argue that the converse is not that case. Tensed and first-personal beliefs can replace other indexical beliefs in their roles in locational actions.

If Giulia, whilst at Llyn Teifi, suddenly formed the belief 'the place where I am now is Llyn Teifi' or the belief 'the lake I see before me now is Llyn Teifi', then she would become motivated to head west to get home. In this way a tensed first-personal belief is sufficient to play the role of an objective indexical belief or a spatial indexical belief. Plausibly, the same can be said of other indexicals too so that, for example, a use of 'you' might be replaced by a use of 'the person I am now addressing' and so on.

This does not mean that tensed, first-personal, or tensed first-personal language can translate all other indexical language. It merely shows that the former can provide awareness of the same reasons to act that the latter language can. It can play the same role in rational action.[20]

I will not try to settle whether tensed first-personal indexicals can translate all other indexicals, but I will note that an argument given by Castaneda (1967) to show that this is not the case fails. Castaneda has argued that different indexicals cannot be taken to mean the same thing. 'The place where I am now' and 'here', for example, can be shown to have different meaning by considering the sentence 'the place where I am now is here'. This sentence appears to be contingent whereas the sentence 'the place where I am now is the place where I am now' appears to be necessary. Therefore, the former has a different meaning to the latter. Be that as it may, these appearances are misleading and the argument fails. For example, if I utter "The place where I am now is the place where I am now" whilst moving quickly, I will utter the first 'the place where I am now' at a different location to the second 'the place where I am now' and therefore my utterance will be false. It follows that the sentence 'the place where I am now is the place where I am now' isn't necessarily true, and hence, that it doesn't necessarily differ in meaning from the sentence 'the place where I am now is here' which isn't necessarily true. Still, I am not committed to the claim that tensed first-personal indexicals can translate all others and don't mean to suggest that they can.

3.5 Mellor's Tenseless Account of Essential Tensed Beliefs

Mellor, as we saw in the last chapter, adopts a tenseless theory. Despite this he admits that tensed beliefs are essential for timely actions. It would be natural to think that tensed beliefs are required for actions because through them an agent is aware of something that the agent is not aware of through tenseless beliefs, specifically reasons for those actions. In order to undermine this idea Mellor (1981, ch. 5; 1998, ch. 6.3) provides a rather different account of why tensed beliefs are required for timely actions.

Mellor considers a case like Mike's in which Mellor turns on his radio at 1 o'clock. We are invited to suppose that Mellor does this because he desires to listen to the 1 o'clock news and he believes that if he turns on his radio at 1 o'clock he will be able to listen to the 1 o'clock news. But Mellor realizes that this cannot be the whole story because he had that belief and desire all morning yet only acted at 1 o'clock. So, we need something to explain why he acted then and not earlier. The answer

Mellor proposes, in line with the argument for the essential indexical, is that he acts at 1 o'clock because at that time he forms the belief 'it is 1 o'clock now'.

Mellor believes that our actions are caused by our beliefs (and desires) not by the facts that we are aware of through them. He supports this claim by pointing out that if he had come to believe at some time other than 1 o'clock that 'it is 1 o'clock now', then he would at that time have turned on the radio. So tensed beliefs are necessary for timely actions, even though there are no tensed facts, because the beliefs and not the facts they are an awareness of are what the actions are responsive to. To this extent our actions occur irrespective of the truth-value of the beliefs that they are a response to. Nonetheless, Mellor believes that the truth of our beliefs is relevant to whether or not our actions succeed such that only actions caused by true beliefs are likely to succeed.

Mellor believes, as we saw in Chap. 2, that tensed beliefs can be true despite there being no tensed facts. For example, the tensed belief 'it is 1 o'clock now' will be true if and only if[21] that belief is had at 1 o'clock and if it is it will be made true by one or the other tenseless fact: the belief occurs at 1 o'clock (token-reflexive theory) or 1 o'clock is at 1 o'clock (date theory).[22] The action Mellor makes in turning on the radio at 1 o'clock is an action that will only succeed at a specific time. Thus the belief and action fit nicely together. The belief is only true at a specific time and therefore can only cause the action when it is true at a specific time and hence when the action is likely to succeed. On the other hand, the tenseless beliefs 'Mellor's tensed belief occurs at 1 o'clock' and '1 o'clock is 1 o'clock' are true at all times if true at all and therefore if they caused Mellor's act of turning on the radio at a time when they were true they would not necessarily cause that action at a time at which it would succeed.

Mellor's account of why tensed beliefs are necessary for timely actions thus runs as follows. Actions which are caused by true beliefs are more likely to succeed than actions which are caused by false beliefs. A successful timely action must occur at a specific time. A tensed belief is a belief that is only true at a specific time and hence a belief such that if it causes an action when it is true, it will cause an action at a specific time. Therefore, a tensed belief is required for a successful timely action. A true tenseless belief can be true at any time and therefore could cause an action at any time and therefore is not appropriate for causing an action at a specific time.

3.6 Inadequacies of Mellor's Tenseless Account

We should be careful how we interpret the idea that tensed beliefs are only true at specific times. Mellor in fact says that tensed beliefs, like tenseless ones, can have an unchanging truth-value.[23] In order to make this clear I will consider Mellor's own distinction between a token and a type belief. A belief type might be embodied in numerous different belief tokens. Thus if Mellor and I both believe 'it is 1pm now', we will have different token beliefs but these beliefs will be of the same type.

When talking of belief tokens Mellor in fact says that they will always have an unchanging truth-value if they are of a short duration. His point is that if I yesterday at 1 o'clock believed 'it is 1 o'clock now', my belief would then have been true. Further, it remains the case that that belief was true even though it is not now 1 o'clock. In short, the truth-value of the token remains constant despite the token being tensed. In a similar way a tenseless token will have an unchanging truth-value: if that belief token is true at any time, then it will be true at every time.

Nonetheless, there is a sense in which a brief true tensed token will only be true at a specific time. This is the sense in which the token will only have properties when it exists. So, if it only exists at a specific time, it will only be true at that specific time. However, this does not distinguish a brief true tensed token from a brief true tenseless token. A brief tenseless token will also only be true when it exists and hence at a specific time.

A drawn out tensed belief token will only be true at a specific time in a different sense. For example, if one believes 'it is 1pm now' for three hours starting at midday, the belief will be false for a while, then true, and then false again. However, such drawn out tokens are not at issue precisely because one does not require a drawn out tensed token to act. In fact, such drawn out tensed tokens are bound to be wrong at times and are therefore quite rightly rare. (A token like 'it is now 2015' might be drawn out for a year, but just for a year and hence not drawn out in a way which makes it only true at a specific time in the sense being considered in this paragraph.)

The case with belief types is slightly different. A tensed belief type might have some tokens which are true and some which are false. Whereas, a tenseless type will have tokens all of which have the same truth-value. But this is not to say that a tensed belief type changes truth-value. Rather, it is to note that a tensed belief type does not have a truth-value at all. (In fact it cannot have a truth-value because Mellor's theory of truth specifically refers to a time of occurrence and a type does not have a time of occurrence.) Moreover, Mellor is concerned with the causes of actions and it is only belief tokens, not types, which can be causes. So we must focus on tokens and when we do we see that tensed and tenseless tokens are, or can be, alike in their nature as truth-bearers. So, this cannot be what explains the need for tensed beliefs. Tensed and tenseless tokens can both be true only at a specific time. Therefore, the need for a belief only true at a specific time cannot be what makes a tensed belief essential for a rational timely action.

Mellor might have a true[24] tenseless belief such as 'God is love' only at 1 o'clock[25] and this might cause him to turn the radio on. If this happens all of Mellor's criteria are fulfilled. The action is caused by a true belief, moreover, a belief that is only true at a specific time and which causes an action that will only succeed when that belief is true. This possibility therefore shows that Mellor's criteria do not explain why tensed beliefs are necessary for timely actions.

This example might appear unfair to Mellor because God's being love appears to be of little relevance to Mellor's turning on the radio to hear the news. But, the notion of relevance doesn't actually support Mellor's case. It is not clear that Mellor can show that the tensed belief that he offers—that 'it is 1 o'clock now'—has any particular relevance to the action.

Now, I do believe that the fact that it is 1 o'clock provides one with a reason to act to turn on the radio if one has a reason to hear the 1 o'clock news. Because of this I think that the belief that 'it is 1 o'clock now' also has relevance to the action of turning on the radio at 1 o'clock. But Mellor, because he does not believe in the tensed fact, cannot say this.

The only fact that exists and is given by the tensed belief for Mellor is its truth-maker. This truth-maker is either the fact that *the belief occurs at 1 o'clock* or the fact that *1 o'clock is 1 o'clock*. But these facts are quite irrelevant to the action of turning on the radio at 1 o'clock. The latter fact is a tautology and whilst loosely relevant it is not something that helps increase the particular likelihood of the success of Mellor's action. At any rate, it clearly doesn't increase the likelihood of success any more than the fact that *2pm is 2pm* or the fact that *3pm is 3pm*, etcetera. So Mellor seems to have no more need to believe 'it is 1 o'clock now' than to believe 'it is 2 o'clock now' or 'it is 3 o'clock now', etcetera.

Whether or not some beliefs occur at 1 o'clock also appears to have little relevance to the success of Mellor's action. All sorts of beliefs might occur or not occur at 1 o'clock and Mellor's action of turning on the radio will be completely unaffected. What matters to the success of the action is the state of the radio not the state of Mellor's or anyone else's minds.

It is no good to say that the occurrence of Mellor's belief at 1 o'clock has relevance to the success of his action because that belief causes the action, because at this stage we are trying to uncover why the belief must cause the action. To say that Mellor's tensed belief must cause the action because it is relevant, and it is relevant because it causes the action, is no explanation at all. So, whilst it seems appropriate to insist that the belief that explains Mellor's action must be relevant to the success of his action, it is not obvious that this is something that Mellor can say.[26]

Davidson (1963/2001a, 1978/2001b), who like Mellor believes that actions are caused by belief/desire complexes, does go a little further. He says that these belief/desire complexes must rationalize the action. It might seem that the tensed belief 'it is 1 o'clock now' will rationalize the action of turning on the radio in a way that the tenseless belief 'God is love' will not. However, this is now to give a quite different account for the need for tensed beliefs, namely, that they can rationalize actions which tenseless beliefs cannot. As it happens I think this is right. But, it is far from what Mellor says. Further, it is not clear that Davidson's position can explain why we must have tensed beliefs because we can have third-personal and tenseless practical syllogisms that rationalize an action in a tenseless way. For example: on the 12 of August Mellor desires to hear the 1 o'clock news; it is necessary that if Mellor hears the 1 o'clock news on the 12th of August Mellor turns on the radio at 1 o'clock on the 12th of August; therefore, it ought to be the case that Mellor turns on the radio at 1 o'clock on the 12th of August.[27]

Rationalization in the Davidosnian sense doesn't explain the need for tensed beliefs. Nevertheless, rationality is crucial. The tensed beliefs do play a role in rational actions that tenseless ones cannot. One natural way to understand why this is is that tensed beliefs provide us with reasons that tenseless ones do not and a rational action is one which is responsive to the appropriate reasons. In addition, it is natural to take the reasons provided to be the objects of these beliefs not the beliefs themselves. If I go to the shop because I believe it sells milk and it does not in fact sell milk, then when I arrive at the shop and realize my belief is mistaken I will also realize that I had no reason to go there after all. Our beliefs do come between our reasons and our actions because we must be aware of our reasons to act. Sometimes we can have mistaken beliefs and so our actions can fall out of line with the reasons that there are to perform them. But this does not mean that our actions are not generally responsive to reasons as states of the world.

Mellor, as a tenseless theorist, cannot tell a story like this. Instead he must refer to a difference between tensed and tenseless beliefs which is independent of their objects (the facts they pick out). He tries to do this with reference to the fact that a tensed belief unlike a tenseless one is true at a specific time. But, a tenseless belief can also be true at a specific time. Mellor cannot simply say the tensed beliefs provide reasons not provided by the tenseless ones, because we need an account of why this is. Such an account cannot be based on the idea of rationalizing an action because the natural way to understand this notion of rationalizing does not prevent one from constructing a tenseless rationalization. If Mellor leaves it as a brute fact that tensed beliefs provide reasons tenseless ones cannot, then he offers an account which is poor in comparison to the tensed theory's account. The tensed theory can explain why a tensed belief provides a reason that a tenseless one does not by referring to the object of that belief. Moreover, if the tensed theory is right that reasons are the objects of beliefs, not the beliefs themselves, then the tenseless theory cannot accommodate this in its theory because it can't distinguish between the facts picked out by tensed and tenseless beliefs nor hence can it say that one provides reasons the other does not.

There is a move open to the tenseless theory that Mellor does not emphasize but that appears to have relevance here. Mellor could have argued that tensed and tenseless beliefs differ, not because only the former can be true at a specific time, but because only the former can be known to be true at a specific time. In order to make this clear let us distinguish a number of beliefs:

- *B4*, '*it is 1 o'clock now*'. This is the tensed belief that Mellor acts on.
- *B3*, '*belief b*[28] *occurs at 1 o'clock*' and '*1 o'clock is 1 o'clock*'. These are two tenseless beliefs which share truth-makers with B4 according to the tenseless theory (one tenseless belief for each of the two possible candidate truth-makers). I will treat them together as belief B3 because the arguments apply to them equally and so treating them together simplifies matters.
- *B2*, '*if I act on my tenseless belief 'belief b occurs at 1 o'clock' when it is true, then I will succeed in hearing the 1 o'clock news*' and '*if I act on my tenseless belief '1 o'clock is 1 o'clock' when it is true, then I will succeed in hearing the 1 o'clock news*'. This is a pair of beliefs about Mellor's tenseless belief B3. Again I treat these two together as B2 to simplify matters.
- *B1*: '*if I act on my belief 'it is 1 o'clock now' when it is true I will succeed in hearing the 1 o'clock news*'. This is a belief about Mellor's tensed belief B4.

The thought is that Mellor could know B1 all morning. However, Mellor could not know B2 all morning. Therefore, Mellor must act on a tensed belief like B4 and not a tenseless one like B3.

As it stands, this argument is mistaken. Mellor could believe B2 all morning. Be that as it may, there is something awkward in this because believing B2, whilst distinct from believing B3, is nevertheless closely linked to it. It is natural to move from the former belief to the latter belief insofar as if one knows that one can in the future have a tenseless belief and it be true, it is natural to hence adopt that tenseless belief in the present. No similar link holds in the tensed case. One can believe that a tensed belief will be true at some time in the future without being in anyway moved to therefore adopt that tensed belief in the present. If one did form the tenseless belief B3 upon forming belief B2 early in the morning, then B2 will no longer be correct. This is the sense in which B2 could not be believed all morning.

This picks out a genuine asymmetry between tensed and tenseless beliefs, such as B4 and B3. Nevertheless, the asymmetry only has relevance to Mellor's action of turning on the radio if Mellor must have knowledge about when the beliefs that cause him to act are true in advance of his acting. Specifically, it must be the case that a person must know in advance of acting because of a belief (i.e. B4 or B3) that that belief, when true, will lead the person to act successfully. That is, one must know something like B1 or B2. However, it is not clear that Mellor could suppose that such knowledge is required for the action at issue or even that such knowledge would aid his case.

Firstly, when one acts in turning on the radio at 1 o'clock the phenomenology of the situation does not include a higher order belief such as B1 or B2. Secondly, if we suppose that one must have knowledge like this (i.e. B1 or B2), then it no longer seems plausible that one must also know that the belief one acts on (i.e. B3 or B4) will be true when one has it. If I act on forming a belief because I know that in so doing I will succeed, then I do not in addition to knowing this also have to know that the belief I have formed will be true when I act on it. To know that acting on it will be successful all I need to know is that it occurs at a certain time and it is possible to know that one will have a belief at a certain time without knowing that it will be true at that time, because the truthmaker of a belief is not the only thing that can bring it about. (I might know that I am off to see a hypnotist at 1 o'clock and that she will convince me to believe something false at 1 o'clock. Then I will know that if I act when I have that belief, then I will act in a successful timely manner. I will know this without having to know that the belief will be true when I have it or act on it.) If one needn't know that the belief one ought to act on will be true, then the asymmetry between the tensed and tenseless cases evaporates. It was only knowing the tenseless belief (i.e. B3) would be true in the future that might encourage one to form the tenseless belief in the present (and hence render B2 false). Moreover, why would the object of beliefs such as B1 and B2 have to be beliefs that occur at a certain time rather than some other timely event that is plausibly easier to predict? It would be sufficient for an agent to know simply that if he or she acts when a clock says 1 o'clock, then he or she will likely succeed in acting. Once we have the higher order belief the lower order one is no longer necessary. Or more accurately, once we have the belief about a timely event it is no longer necessary that the timely event is the occurrence of a belief.

Thirdly, if Mellor must believe B1 or B2 in order to perform a rational timely action of turning on the radio, then this implies that this knowledge plays a role in the rationality of his action. But it is not clear how Mellor can account for this. The natural way to understand why one needs knowledge such as B1 and B2 to be rational is to take beliefs B3 or B4 to be or to capture reasons to act. But we have just seen that Mellor might struggle to justify this idea on a tenseless framework.

Thus there is an asymmetry between the case of knowing about one's tensed beliefs and knowing about one's tenseless beliefs. Nevertheless, this cannot account for our need for tensed beliefs because knowledge about one's tensed or tenseless beliefs is not itself required in the circumstances in which tensed beliefs are needed. The asymmetry is irrelevant to the case at hand.[29]

3.7 Summary

In this chapter we have seen that Perry's argument for the essential indexical shows that tensed beliefs are essential for rational timely actions. Perry's initial argument specifically concerned a timely and rational action, not just any old doing, and plausibly applies to all actions of this type.

This fact appears to lend initial credence to the tensed theory. It implies that when one acts on a tensed belief this is rational in a way that acting on a similar tenseless belief would not be. And, a natural way to understand this is that one is aware of reasons for one's action, and that one acts for, through one's tensed belief that one is not through one's tenseless belief.

The tenseless theory cannot accept such an account. Mellor, from the stand point of a tenseless theory, offers an alternative account of the need for tensed beliefs based on the fact that tensed beliefs, unlike tenseless ones, can be held and true at a specific time.

A consideration of Mellor's account found it to be inadequate because the asymmetry he alluded to was not a genuine asymmetry. Tenseless beliefs can also be held and true just at specific times. Mellor's account could be made plausible if a tensed belief is a reason to act whilst a similar tenseless one is not. However, it is not clear that the tenseless theory can justify such a claim. Mellor's account could also be made more plausible if one had to know in advance of acting that the belief one would act on in acting would be true at the time one believed it and would also make one's act likely to succeed (e.g. B1 or B2). However, again, it is not clear that Mellor can justify such a claim because it is not clear that such a belief is rationally demanded and nor is it clear that if such a belief were necessary that it would make acting on a tensed belief essential to a rational action.

As with Chap. 2, we have seen that a clear understanding of the rationality of our actions and our emotions can play an important role in the debate between the tensed and tenseless theories. In the subsequent chapters I will therefore turn to investigating the nature of this rationality.

Notes

1. In Chap. 9 we will see that matters are a little more complicated than this because the reason an emotion is felt for must also justify having that emotion.
2. Of course, all words can change their meaning or referent over time but such a protracted change of use is not what is at issue here. Rather, indexicals can change their referent without, we might say, changing their use or meaning.
3. Cf. Kamp (1972), Smith (1990, 1993), Prior (1968), and Salmon (1986, 34–40).

4. I am not alone in treating them alike in these situations. Cf. e.g. Reichenbach (1948) and Perry (1997).
5. When I say this I do not mean that the person at issue would in fact be willing to say anything at all. The person might be in hiding. Rather, the point is that s/he would ideally express it by saying "I am making a mess" or s/he would express it that way if s/he were willing to honestly express her/his belief at all. Thus we can say that the belief and the utterance share content (in a broad sense). One might consider the converse of this, such that, when Tom says (honestly etc.) "I am X" or "I believe I am X", we take this to inform us that Tom has the belief 'I am X'. Cf. Kripke's disquotational principle in Kripke (1979).
6. It is worth noting that the mirror is not essential to such a point, e.g., when a number of people put their hands together in the centre of a circle it is possible for people to fail to recognize their hand as their own. Cf. Boer and Lycan (1980).
7. Castaneda (1966) makes a very similar point, though concerning 'he himself' rather than 'I'. This is merely a shift of perspective because the use of 'he himself' considered is essentially that of attributing first-personal beliefs to another. E.g., we say Tom has the belief 'I am *F*' or similarly Tom believes he himself to be *F*. Cf. Castaneda (1967, 1968).
8. Cf. Alvarez (2005, 45).
9. Cf. Anscombe (1957/2000, sect. 5, p. 9).
10. Though, we might well deny this. Cf. Lowe (2008), Dancy (2000), and Alvarez (2007).
11. Cf. e.g. Dancy (2000).
12. Cf. Wason (1966) for a commonly discussed example of widespread irrationality. Cf. Evans and Over (1996, 7) for an account of irrationality which matches my description here of acting for bad reasons.
13. If one was inventive s/he could create an example where the states were seen directly rather than simply on a screen, but I'll spare you this gore.
14. To talk of this 'presence of experience' is not to presuppose the tensed theory, rather the phrase is simply used to capture the form of awareness that Mike must have to be moved to act.
15. According to many theories of demonstratives an agent must recognize the referent of a demonstrative to be playing a particular role in a context in order to understand the use of the demonstrative. To this extent perception might commonly be involved in demonstrative reference. E.g., some say you must recognize the referent as the thing demonstrated (e.g. Kaplan (1989a), Reimer (1991a), and perhaps Salmon (2002)), others

say that you must recognize it as the most salient thing in the context (e.g. Wettstein (1984), Reimer (1991b)), or as the object of the speaker's intention in the context (e.g. Kaplan (1989b), Bach (1992a, 1992b)), or as the thing in a particular spatiotemporal relation to the speaker in the context (e.g. McGinn (1981)). (These distinctions are not necessarily exclusive of one another and are rather rough with some proponents falling into more than one group and proponents within one group differing in detail).

16. One is tempted to say simply 'one can know of what one demonstratively refers to that it is present', but my opponent of course is not allowed to say this and if s/he doesn't it is not even clear this argument is at all potent because Mike might well know a belief is simultaneous with a perception of it and be unmoved to act.
17. Most of the views outlined in footnote 15 do not entail that one must directly perceive the referent (cf. Kaplan (1970)).
18. Cf. Le Poidevin (2007).
19. There is thus a temptation to see demonstratives as tensed. Broad (1938, pt. I ch., XXXV, sect. 2.24, p. 305–307) makes this implication by noting that the copula following a demonstrative appears to be tensed.
20. If translation were possible, then this would mean that all indexical language beside first-personal language is in fact tensed language. But, it wouldn't follow from the fact that some tensed first-personal language can translate all other indexical language that non-tensed non-first-personal language can translate all tensed and all first personal language.
21. Actually Mellor avoids stating this as a biconditional but I think this is a more natural reading and this will not affect my argument.
22. Mellor adopts a token-reflexive theory in (1981) and a date theory in (1998). His shift in view is due to arguments that Smith (e.g. (1993)) has given against the token-reflexive theory. I think that Smith's arguments are damning of the token-reflexive theory, but I have no need to rely on them here.
23. E.g. Mellor (1981, ch. 6; 1998, ch. 7.4).
24. Let us suppose.
25. It is quite possible to have a tenseless belief for only a short duration. In an unusual moment of inspiration I might consider God's nature, decide that he is love, and then quickly abandon that idea. Alternatively, I might briefly think to myself that I have thought *x* at 1 o'clock, then abandon

that idea because I was not actually sure whether it was thought *x* I was having.

26. The arguments of this paragraph and the two that proceed it show that the reference Mellor makes to success semantics in other writings of his does not enable him to avoid the worry that I have raised here (cf. Mellor (2012)).
27. Cf. McGinn (1979) who quite naturally presents a practical syllogism in a third-personal way.
28. I take 'belief b' to refer to the relevant first order belief provided by the token-reflexive truth-conditions of B4, e.g., 'it is 1pm now'. This mode of reference could be taken to be via a demonstrative, name, or description as required.
29. The idea that beliefs about the beliefs we act on play a role in our actions arises in a recent discussion of these matters offered in Mozersky (2006). My presentation of the idea differs from Mozersky's in a number of ways because Mozersky says that he is just trying to make it coherent how tensed beliefs can play a role in timely actions. We do not find an explanation for the need for tensed beliefs in what he says, rather we get a presentation in which tensed beliefs just are caused by certain things and just do lead to certain inferences and actions (cf. Mozersky (2006, 463)). (Further, one cannot simply infer an explanation form what Mozersky says. Because, e.g., if he is right tensed and tenseless beliefs share content and content is what is at issue in inference, then we have no explanation of why tensed as opposed to tenseless beliefs are needed as premises or arise as conclusions in the inferences present for him in action. But given Mozersky's goal was to make it coherent that tensed beliefs play these roles, rather than to explain why they must play them, this should not be surprising.)

References

Alvarez, M. (2005). Agents, Actions and Reasons. *Philosophical Books, 46*, 45–58.

Alvarez, M. (2007). The Causalism/Anti-Causalism Debate in the Theory of Action: What It is and Why It Matters. In A. Leist (Ed.), *Action in Context* (pp. 103–123). New York: de Gruyter.

Anscombe, G. E. M. (1957/2000). *Intention*. London: Harvard University Press.

Bach, K. (1992a). Intentions and Demonstrations. *Analysis, 52*, 140–146.

Bach, K. (1992b). Paving the Road to Reference. *Philosophical Studies, 67*, 295–300.

Boer, S. E., & Lycan, W. G. (1980). Who, Me? *Philosophical Review, 89*, 427–466.

Broad, C. D. (1938). *An Examination of McTaggart's Philosophy* (Vol. II). London: Cambridge University Press.

Castaneda, H.-N. (1966). He': A Study in the Logic of Self-Consciousness. *Ratio, 8*, 130–157.

Castaneda, H.-N. (1967). Indicators and Quasi-Indicators. *American Philosophical Quarterly, 4*, 85–100.

Castaneda, H.-N. (1968). On the Logic of Attributions of Self-Knowledge to Others. *Journal of Philosophy, 65*, 439–456.

Dancy, J. (2000). *Practical Reality*. Oxford: Oxford University Press.

Davidson, D. (1963/2001a). Actions, Reasons, and Causes. In *Essays on Actions and Events* (pp. 3–19). Oxford: Oxford University Press.

Davidson, D. (1978/2001b). Intending. In *Essays on Actions and Events* (pp. 83–102). Oxford: Oxford University Press.

Evans, J. S. B. T., & Over, D. E. (1996). *Rationality and Reasoning*. Hove: Psychology Press.

Kamp, H. (1972). Formal Properties of 'Now'. *Theoria, 37*, 227–273.

Kaplan, D. (1970). Dthat. *Syntax and Semantics, 9*, 221–243.

Kaplan, D. (1989a). Demonstratives. In J. Almog, J. Perry, & H. Wettstein (Eds.), *Themes from Kaplan* (pp. 481–563). Oxford: Oxford University Press.

Kaplan, D. (1989b). Afterthoughts. In J. Almog, J. Perry, & H. Wettstein (Eds.), *Themes from Kaplan* (pp. 565–614). Oxford: Oxford University Press.

Kripke, S. (1979). A Puzzle about Belief. In A. Margrlit (Ed.), *Meaning and Use* (pp. 239–283). Dordrecht: Reidel.

Le Poidevin, R. (2007). *The Images of Time*. Oxford: Oxford University Press.

Lowe, E. J. (2008). *Personal Agency*. Oxford: Oxford University Press.

McGinn, C. (1979). Action and Its Explanation. In N. Bolton (Ed.), *Philosophical Problems in Psychology* (pp. 20–42). London: Methuen.

McGinn, C. (1981). The Mechanisms of Reference. *Synthese, 49*, 157–186.

Mellor, D. H. (1981). *Real Time*. Cambridge: Cambridge University Press.

Mellor, D. H. (1998). *Real Time II*. London: Routledge.

Mellor, D. H. (2012). *Mind, Meaning, and Reality*. Oxford: Oxford University Press.

Mozersky, J. M. (2006). A Tenseless Account of the Presence of Experience. *Philosophical Studies, 129*(3), 441–467.

Perry, J. (1979). The Problem of the Essential Indexical. *Nous, 13*, 3–21.

Perry, J. (1997). Indexicals and Demonstratives. In B. Hale, & C. Wright (Eds.), *A Companion to the Philosophy of Language* (pp. 586–612). Oxford: Blackwell.

Prior, A. N. (1959). Thank Goodness That's Over. *Philosophy, 34*, 12–17.

Prior, A. N. (1968). Now. *Nous, 2*, 101–119.

Reichenbach, H. (1948). *Elements of Symbolic Logic*. New York: Macmillan.

Reimer, M. (1991a). Demonstratives, Demonstrations, and Demonstrata. *Philosophical Studies, 63*, 187–202.

Reimer, M. (1991b). Do Demonstrations Have Semantic Significance? *Analysis, 51*, 177–183.

Russell, B. (1940/1995). Egocentric Particulars. In *An Inquiry into Meaning and Truth* (pp. 108–115). London: Routledge.

Salmon, N. V. (1986). *Frege's Puzzle*. Cambridge, MA: Massachusetts Institute of Technology Press.

Salmon, N. V. (2002). Demonstrating and Necessity. *The Philosophical Review, 111*, 497–537.

Smith, Q. (1990). Temporal Indexicals. *Erkenntnis, 32*, 5–25.

Smith, Q. (1993). *Language and Time*. Oxford: Oxford University Press.

Wason, P. C. (1966). Reasoning. In B. M. Foss (Ed.), *New Horizons in Psychology* (Vol. I, pp. 135–152). Harmandsworth: Penguin.

Wettstein, H. (1984). How to Bridge the Gap Between Meaning and Reference. *Synthese, 84*, 63–84.

Part II

Rationality

4

Reasons

In this chapter I will examine the nature of reasons. We saw in Chaps. 2 and 3 that in order to progress certain debates about the nature of time we need to get clear about the rationality of actions and the appropriateness of emotions. The rationality of an action is tightly bound up with reasons for that action, just as the appropriateness of an emotion is bound with the reasons that emotion is felt for. Whether or not we reduce this rationality and this appropriateness to reasons it is clear that we cannot fully assess the former without considering the nature of the latter. (In my discussion I will focus on actions because this is the focus common in the literature, but what I say is applicable to the case of emotions as I will make clear in Chap. 9.)

I take reasons to be facts. For example, the fact that a car is approaching me is a good reason for me to jump out of the road. Further, if I spot the car, I will jump for this reason. I will begin this chapter by introducing terminology popular in the philosophy of reason and using it to clarify my view. It will become clear that the points I make in the second and third sentences of this paragraph require independent defence. I will promote the idea that reasons are facts and will defend it from numerous arguments.

Olley (F.O.C.H.) Pearson, *Rationality, Time, and Self*,
https://doi.org/10.1007/978-3-319-71973-3_4

4.1 Theories & Terms

We use the notion of a reason in a variety of ways. Alvarez (2009) picks out three. Firstly, we can speak of reasons that justify actions: Tom's mess-making justifies his stopping. Secondly, we can speak of the reason someone acted for or what motivated him or her: Tom might stop for the reason that he is making a mess. Thirdly, we can speak of things that explain an action as the reasons for that action: we understand Tom's stopping through being made aware that he was making a mess.

In line with these distinctions we can speak of justificatory, motivational, and explanatory reasons respectively. However, as Alvarez is careful to stress, this need not mean that we are actually dealing with three distinct types of reasons or even three distinct reasons. It could well be that one and the same fact fulfils all three roles and hence is at once a justifying, motivating, and explanatory reason. For example, Tom's making a mess justifies his stopping, motivates him to stop, and can explain his stopping. On the other hand, it is possible for the three to diverge. A fact might justify Tom's act, but if he is unaware of it it will not motivate his action and might not explain it. Tom might be motivated to act by a fact that does not actually justify his action. Tom's action might be explained by a fact that neither justifies nor motivates his action. Similarly, a single fact might justify one act, motivate another, and explain a third. The difference between the types of reason is not a difference in the nature of the reason but rather in the role the reason plays. A fact is a justificatory reason if it justifies an action, it is a motivational reason if it motivates an action, and an explanatory reason if it is used to explain an action. With this possibility of coincidence and divergence clear I will go on to speak of three types of reason for simplicity though the distinction is centrally one of three roles.

For simplicity I will refer to justificatory reasons as *j-reasons*, motivational reasons as *m-reasons*, and explanatory reasons as *e-reasons*. (When I do not need to delineate a single role I will often drop the prefix.)

For each type of reason there can be a theory about what sort of things those reasons are. I have been speaking of all three as facts and thus I allow that all three can coincide, and I will defend this view below. However, some disagree and take the different types of reason to fall into different ontological categories hence denying coincidence. For example, Wallace[1] takes j-reasons to be facts but he thinks that m-reasons are psychological states. Tom is justified in acting by the fact that he is making a mess but he is motivated to act by his believing that he is making a mess and his desiring not to. I will refer to the view that reasons are psychological states as *psychologism about reasons*. More specifically, I shall speak of the view that j-reasons are psychological states as *psychologism about j-reasons*, the view that m-reasons are psychological states as *psychologism about m-reasons*, and so on for e-reasons. I will refer to the view that reasons are facts analogously as *non-psychologism about j-reasons*, *non-psychologism about m-reasons*, and *non-psychologism about e-reasons*.

Psychologism about j-reasons must be distinguished from *internalism* about j-reasons.[2] Internalism is the view that one's j-reasons are in some way delimited by one's motivation or potential motivation. For example, Williams (1979) has defended a view according to which one only has a j-reason to *F*, if one has a desire that would be satisfied by *F*ing or could be brought to have such a desire through rational deliberation after having been made better aware of the situation. Internalism thus places a condition on any j-reason, however, it does not, as psychologism does, impose an ontological form on those reasons. One could adopt an internalism like Williams' and adopt either non-psychologism or psychologism. Tom's act could be justified by the fact that he is making a mess even if it is only so justified as long as Tom could (in the relevant way) be motivated by that fact. *Externalism* in this context is the denial of internalism.[3]

It is important to distinguish between a belief—a mental state[4]—and what is believed—a state of the world the believer is aware of. The term 'belief' can be used to refer to either of these and thus it has an act/object ambiguity. A similar act/object ambiguity occurs for the term 'desire' which might be used variously to refer to the mental state of desiring or to the object of such affections.[5] Psychologism is the view that j-reasons are mental states and in this sense the view that j-reasons are beliefs and desires. We must therefore be careful. I will use the expressions 'belief' and 'desire' to speak of mental states unless I make it clear otherwise. The terms 'belief' and 'desire' are commonly used in the literature in an umbrella sense. 'Belief' picks out a number of cognitive states so that if one believes *b*, one might in fact doubt *b*, assume *b*, know *b* etcetera. 'Desire' is used to capture a number of pro-attitudes so if one desires *d*, one might in fact want *d*, be inclined towards *d*, etcetera. I will say more about this in the following chapter but for the time being I will also adopt this umbrella use.

With these distinctions in the open it is possible to delineate a number of different views crisscrossing these distinctions. The view that I wish to defend is non-psychologism about j- and m-reasons. I will not focus on e-reasons but what I will argue is compatible with facts[6] and beliefs being e-reasons. I will stay largely quiet on the issue of internalism versus externalism except for the following remarks.

There are a number of different forms that internalist theories could take that embody the general idea that reasons are in some sense related to motivation. Chapter 3 raised the question of why indexical beliefs might be necessary for certain rational actions. If one adopts a particularly extreme form of internalism one will gain some opportunity to provide a simple answer to this question. For example, one might adopt an internalist theory according to which one has a j-reason only so long as one can be motivated to act whilst having an indexical belief. If one then insisted that one can only act rationally if one has j-reasons, it would follow that Tom must have an indexical belief to act rationally because without one nothing he did could be rational. However, such an extreme form of internalism appears to be very difficult to justify. It certainly doesn't strike one as intuitive. Rather, as long as Tom is making a mess he has a reason to act whether or not he has an indexical belief. In this regard it is telling that the forms of internalism commonly defended in the literature are far less extreme and hence more plausible.[7] Of these more common forms of internalism Williams' is perhaps one of the more hardline.[8] However, I think it is clear that an internalism of this form does not explain why indexical beliefs are needed for rational actions. The fact that Tom is making a mess could qualify as a reason without Tom having an indexical belief whether or not we adopt this internalism. Hence, for matters of simplification I shall put the internalism/externalism debate to one side on the assumption that any plausible or common form of internalism (and hence externalism) will not prevent the questions of the essential indexical from arising.

Occasionally a distinction is made between *conclusive* and *pro tanto* reasons. If one has a conclusive reason to do something then one ought rationally to do it. On the other hand, someone might have a *pro tanto* reason to do something but also have stronger *pro tanto* reasons not to do that thing so that all things considered he or she ought rationally not to do it. *Pro tanto* reasons can be weighed against one another and having a *pro tanto* reason to do something does not mean that all things considered you ought rationally to do that thing or that that thing is the most reasonable thing to do. I think this raises an important issue and it should be understood that when I say that someone has a reason to do something I do not mean that all things considered that person ought to do that thing or that that thing is the most reasonable thing to do. In short, I will generally be discussing *pro tanto* reasons unless I make clear otherwise.

4.2 Justificatory Reasons

The greatest adversary to non-psychologism about j-reasons is psychologism about j-reasons.[9] (From here on in this section I will often simply speak of non-psychologism or psychologism. In doing so I mean to speak of non-psychologism and psychologism about j-reasons.) When the two are considered shoulder to shoulder I think it is very clear that non-psychologism is the better theory. For example, Tom's j-reason to stop and rearrange his shopping appears to be the fact that he is making a mess. It is not his belief that he is (nor his belief in anything different, nor anyone else's belief). There is a j-reason for him to stop whether or not he has this belief. The fact that he is making a mess does not stop being problematic just because he does not believe he is and it does not stop being a j-reason for him to act to stop making a mess.

People consider facts when deliberating about what they ought to do or to have done. I see the fact that it is raining as a j-reason to take an umbrella and the fact that the only umbrella available belongs to someone else as a j-reason not to take an umbrella. I do not consider my beliefs but rather what they are beliefs of—facts—in order to make my decision.[10]

Kim (1998) has pointed out that if I were to consider my beliefs and desires this would move me from an act of deliberation to one of prediction. In considering that I believe it is raining, that I believe the only available umbrella belongs to someone else, and that I desire not to use other people's belongings I can predict that I will not take the umbrella in the same way that another person may predict my behaviour through knowing my mental states. But this conclusion will not tell me what I ought to do. It will not be the conclusion of an act of deliberation considering my j-reasons for acting and it will leave me still with having to make a decision about what to do. (Predicting what I am likely to do and deciding what I ought to do are clearly different as I can decide that I ought to try to avoid the predicted action.) In order to make this decision about what to do I need to deliberate about the facts that obtain.

Suppose that I lend someone twenty pounds after he tells me that he needs it to pay his phone bill.[11] If I later discover that he was lying and that he just wanted to buy some alcohol, then I will feel that I in fact had no j-reason to lend him money and that I acted for no good reason. I think to myself that I shouldn't have given him the cash. This implies that it was the supposed fact that he needed to pay a bill that was what I took to be a j-reason for my lending the money. If the j-reason had merely been my belief that he needed the money to pay a bill, then I oughtn't to suppose that the act was unjustified upon discovering the falsity of the belief. Discovering the belief was false does nothing to undermine the past existence of the belief nor hence the potential for the belief to have justified the act. I learn that my act was unjustified because I learn that the fact I took to be the j-reason did not obtain.

For these considerations I take it that non-psychologism about j-reasons ought to be our default position. Any other view would have to unsettle non-psychologism before it could be accepted.[12]

Psychologism is driven by cases such as the following. If Si's mother, Jo, is unwell, then Si has a j-reason to call around on his way home to see if there is anything he can do for her. Here the j-reason is the fact that Jo is unwell. The opponents of non-psychologism, however, point out that even if Jo is in fact quite well, if Si believes she is unwell this seems to give grounds for saying that he ought to call around to see her. In this case our original j-reason no longer occurs so if Si ought to act in this way it must be the result of another j-reason, presumably his believing Jo unwell. We thus appear to have grounds for taking a j-reason to be a psychological state, contra non-psychologism.

However, psychologism or the thought that beliefs/desires provide j-reasons must be mistaken. One can easily imagine an act so heinous that no one ever ought to perform it, perhaps murdering a child that has a great life ahead of it. Suppose that someone were to suddenly form the desire to commit such a murder or the belief that such a murder was right, and a belief that murdering the child in front of them would be committing such a murder. According to psychologism this person would have a j-reason to commit this murder. The formation of the belief/desire complex brings a j-reason into existence as the j-reason just is that belief/desire complex. But this is obviously false. The murder remains equally unjustified. Forming the belief that an action is right does nothing to make that action more right. One cannot simply bootstrap j-reasons into existence through forming beliefs and/or desires, contra psychologism.[13] In short, psychologism must be wrong. (I will refer to the argument just given against psychologism as the *bootstrapping* argument.) Moreover, further consideration will show that examples which might appear to raise support for psychologism and trouble for non-psychologism in fact fail to do so.

Si appears to be subject to the complex normative requirement (hereon abbreviated to *CNR*) that *one ought, if one believes one's mother unwell, to call around to see her.*[14] Psychologism recognizes such an obligation and assumes that Si has a j-reason to call around to see his mother. But we need not interpret the CNR in this way. Rather, it has been argued that the ought in such obligations must be interpreted to have a wide scope precisely in order to avoid the difficulties of bootstrapping already alluded to.[15] One cannot conclude that Si ought to see his mother from the fact that he has a belief that she is unwell. The obligation cannot be detached in this way. All the CNR requires is that Si either goes to see his mother or loses the belief that she is unwell. He has a requirement to satisfy a disjunction though not a requirement to fulfil either disjunct in particular. Si could satisfy this CNR if he did not believe that his mother is unwell. In fact, given that his belief is mistaken, it would be better if he did not have that belief and did not go to see his mother. This would in no way appear to conflict with the CNR and thus it vindicates the idea that the obligation applies to the disjunction as a whole.

If there is such a disjunctive obligation, then it is surely not provided by Si's belief because it can be satisfied by the absence of that belief. Rather, if there is a j-reason for this disjunctive obligation then it likely comes from a fact, such as the fact that unwell people require help and mothers do a lot for their children.[16] To insist the obligation is not disjunctive or that it stems from a psychological state is to go beyond the example and to simply assume psychologism without defence.

A second but related criticism of non-psychologism concerns advice. One is able to advise someone how to do something, even if one takes there to be no reason to do that thing. Ross might, for example, tell Dave how to make a birthday cake even though he knows that Dave is wrong about the date of Jo's birthday. He has told Dave about his error, but his advice fell on deaf ears. Ross might say "you shouldn't be making a cake, but if you are going to, you should start by getting the ingredients". This is a matter of means end reasoning. Dave desires to make a cake and in order to obtain this end he has to do certain other things including gathering the ingredients. The point of the case is that Dave has no j-reason to make a cake, he has no j-reason to do so even though he desires to, because it is not Jo's birthday.[17] But, despite this it appears that his desire to make a cake has given him a j-reason to do what subserves making a cake. This is because, regardless of whether or not Dave should be making the cake, if Dave went about trying to do it without gathering ingredients then he would be open to rational criticism that he would not be open to if he gathered the ingredients first. Thus, Dave's desire to make the cake appears to provide a j-reason for his gathering ingredients.

I have said that Dave's desire to make the cake is not a j-reason for making that cake. This is a general point supported by the bootstrapping argument already given. Wanting *x* does not give one a reason to have (or do) *x*. Our question now is whether Dave's desire provides a reason for his doing what *subserves* his desired end.[18] The suggestion that it is comes from the idea that Dave would be irrational if he had that desire but did not pursue those means. We have here, again, a CNR: *one ought, if one desires to make a cake, to gather ingredients*. With this in focus we are free to respond to this case just as we did with Si's. Dave is irrational if he fails to meet this CNR and he would do so if he desired to make a cake and he did not gather ingredients. Thus he is open to the criticism of irrationality. However, this is not because he has a j-reason to gather ingredients. As it is he has none. It is only that he has a j-reason (or at least a rational obligation) to satisfy the disjunction 'gather ingredients or do not desire to make a cake'. Because it is not Jo's birthday Dave has no j-reason to make a birthday cake or to desire to, and therefore his not desiring to would be a better way for him to satisfy the CNR.

It is possible for the fan of psychologism to produce numerous examples like Si's and Dave's, and the literature has many. But the pattern of response—citing the CNR and denying detachment—always appears to be appropriate and for this reason I think we can take this discussion to displace a number of other similar cases.[19] Whilst I think these examples form an interesting and provocative line of criticism of non-psychologism, I take my comments here to show that it is inadequate to its purpose. (In Chap. 6 I will say more about CNRs and the nature of their rational import, but my comments here suffice to show that one can accept non-psychologism and accept an obligation on, for example, Si.)

There is a second aspect to Dave's case. Ross has a j-reason to behave in a particular way, for example, to tell Dave that it is not Jo's birthday when approached about the cake. This j-reason looks essentially bound up with Dave's belief and if it just is Dave's belief non-psychologism is faulted. However, non-psychologism has an alternative to this, namely, the fact that Dave has the belief. The j-reason for Ross's act is the fact that Dave believes it is Jo's birthday. Thus non-psychologism is able to account for this situation. Moreover, non-psychologism's account is very plausible. It is not simply the belief, but the possession of the belief at a specific time by a specific person, that is, the fact that Dave has the belief, that provides Ross with a j-reason. Thus, it is clear that non-psychologism is not committed to the implausible idea that j-reasons are isolated from psychological states. Rather, psychological states can play an important role when they are considered as constituents of facts. Moving to the third person perspective makes this distinction more clear.

The final difficulty that I want to consider against non-psychologism concerns the so called fact/value gap. I will consider three arguments that have been given against the idea that values can be facts: the argument from relativity, the argument from derivation, and the argument from queerness. These are of relevance as j-reasons are values, insofar as they are j-reasons that one ought rationally to act in a particular way.

Mackie (1977, pt. 1, ch. 8) raises the argument from relativity focusing on moral values: there are large disagreements between societies geographically and temporally about moral values; if moral values were facts we would not expect this to be the case; therefore moral values are not facts. If one tried to transpose this argument to the case of rational values or j-reasons then one would immediately hit upon a difficulty. It is simply false that there are large disagreements between societies about what is rational. Moreover, it is necessary that this premise is false. In order to understand another person one must assume that person to be by and large rational by one's own standards.[20] Without this assumption it is impossible to get any foothold at all on what that person means or believes. This entails that anyone that did differ radically from oneself about what was rational, or what were j-reasons, would be someone that we could not even understand, let alone describe as disagreeing with us. In the case of rationality the argument from relativity is quite simply impossible. It is likely that not all will agree with me that the fact that Tom is making a mess is a j-reason for him to act. But, in order for two of us to disagree about such a matter we must first agree about a great many more things: we must first agree about a great many more of our j-reasons. Disagreements can only be minor and minor disagreements ought to be expected given that none of us are perfectly rational or reasonable beings.

The argument from derivation occurs in Hume's writings (1739–40/1985, bk. III, pt. I, sect. I). It runs as follows: no statement of value can be derived from statements of fact alone, therefore facts cannot be values. However, if non-psychologism is correct then the premise of this argument is false. More accurately, if non-psychologism is correct then many statements of fact just are statements of value. If the facts at issue are allowed to be values, then the derivation is going to be straight forward. According to non-psychologism a statement of fact that is the conclusion following from other statements of fact might also be a statement of value thereby offering a ready case in which a statement of value is derived from a statement of fact. In short, Hume, for example, allows the derivation of one statement of fact from others, but if non-psychologism is correct then this might just be an example of a derivation of a statement of value from statements of fact.[21]

The argument from derivation has been widely discussed since Hume. For example, Searle (1964) considers an alternative that focuses on the expressions used in the statements. The argument is that one cannot derive a statement using evaluative expressions from statements that do not.[22] This version of the argument meets a similar response to the one just given. But here we can also put more pressure on the issue. For example, a man might have two names but if we do not know that he does we will not be able to derive a statement using the one name from premises which only use the other. On the other hand, if we do know that he has two names then a statement to this effect—a premise involving both names—will appear as a premise in our derivation. So we would still not have an example in which statements involving the one expression—name—can be derived from ones lacking that expression. But we do not take this to show that a man cannot have two names. So even if it were impossible to derive statements involving evaluative expressions from statements that lack them we shouldn't conclude that facts are not values.[23] At the end of the day we are concerned with facts and actions but relations of entailment don't hold between these, they hold between statements which express facts and actions. Failures of derivation or entailment may simply reflect failures of these statements.

Mackie's (1977, pt. 1, ch. 9) argument from queerness is that: if there are objective values then these are objects of a queer sort different from anything else and we would have to know of them through a special faculty. The point being that objective values—like j-reasons as facts—are odd and demand us to be odd.

However, being odd does not render something impossible so non-psychologism is not disproven by this argument, at best it is made less attractive. But non-psychologism can respond more forcefully. The crucial issue is the notion of 'queer' involved. It appears that normal for Mackie is causal nomological or science. The criticism then is that j-reasons are unscientific. This claim is highly dubious. People spend very little of their time doing science, but arguably must respond to j-reasons at every turn. So if anything science is the queer approach to matters. If one is adamant that science will speak the whole and only truth, then one ought to hope that science will encompass these normative facts. Further, in emphasizing the necessary universality of reasoning and rationality one opens space for a law-like and hence science friendly view of j-reasons and rationality.[24]

On the other hand, one can offer an explanation of why values are not fundamental properties of science. Arguably science aims to produce a value neutral description of the world and so will aim to disregard values and hence should not be expected to posit them.[25] The point is we should be careful what Mackie might mean by 'normal' or 'queer'. If by 'normal' Mackie means something like 'value neutral', then we should expect that the normal view makes no reference to values but we should not think that this tells us anything about whether or not values exist.

Moreover, there may be good independent grounds for thinking that reasons and rationality cannot be reduced to a scientific causal system. Davidson's (1963/2001a) causal approach came under fire (from Davidson (1973/2001b) himself at a later date) on the grounds that causal accounts are susceptible to the problem of deviant causal chains. One might be caused by the right reasons, but in the wrong way. (For example, the fact that there is a storm is a j-reason for heading indoors, but actually being forced indoors by the storm is not responding to the j-reason in the right way.) Lowe has further argued that responding to j-reasons cannot be a causal matter because that would undermine the element of free choice essential for rational action.[26] Non-psychologism can thus respond to the argument from queerness by admitting that its objective values are not scientific or causal properties. This would not be an ad hoc answer, but rather one grounded in a thorough understanding of rationality and perhaps science.

To conclude this section, non-psychologism is promoted by our ready consideration of facts as j-reasons. Many arguments against non-psychologism occur in cases where there is a complex normative requirement at play, for example, one ought to *F* if one believes *p*. These CNRs ought to be interpreted in a way that denies detachment. They are a requirement to satisfy a disjunction but not to adopt any one of those disjuncts rather than the other. This removes the pressure to take the psychological state involved to be a j-reason, in fact implying the contrary. In other cases it is a fact constituted in part by a psychological state, rather than the state itself, that is the j-reason (the close relation between these two providing an error theory for psychologism's mistake). Finally, the fact/value gap arguments have their potency stunted when it is noted that rationality must have a universal nature (there can be only one rationality and hence only one set of j-reasons). Further, demands for the possibility of derivation or a particular scientific nature are demands that non-psychologism can regard as inappropriate on independent grounds.

4.3 Motivational Reasons

I now turn to considering the case of m-reasons. (In this section, when I say simply "non-psychologism" or "psychologism" I generally mean non-psychologism or psychologism about m-reasons.) Here the greatest rival to non-psychologism is again psychologism. Our concern now is with the questions: why did Tom act? What motivated him? What was his m-reason for acting? Non-psychologism takes it that Tom's m-reason for acting was the fact that he was making a mess. Psychologism, on the other hand, takes it that Tom's m-reason was his belief that he was making a mess perhaps in combination with a desire not to. Put like this it is clear non-psychologism again has an initial advantage over its rival. Tom would not take himself to be acting for the m-reason that he had that belief and desire. Rather, Tom quite sensibly takes his believing and desiring those things to be no j-reason for his act and hence this is not something that motivates him and therefore is not his m-reason either.

Deliberation again lends credence to non-psychologism. The things one considers when one considers how to act are facts. It is therefore natural to suppose that these facts are what motivate the individual when she or he finally makes her or his decision. To insist that the individual was motivated by something different appears to undermine the very act of deliberating. Deliberation is undermined if one does not act for any of the things that one considers acting for.[27]

Dancy (2000, ch. 5) points out that any account of m-reasons must meet two criteria: EC, it must make it clear how a j-reason can contribute to the explanation of an action done for that reason; NC, it must make it clear how one can act for good j-reasons. To deny NC would be to assert that no one ever acts for good j-reasons, and perhaps hence that no-one is rational. Given non-psychologism about j-reasons it is hard to see how anything other than non-psychologism about m-reasons can meet the latter of these constraints and thus non-psychologism about m-reasons gains credence.

It has been argued that one should not take NC literally and that enough connection is maintained with normative reasons if they are, for example, the content of beliefs that are m-reasons.[28] However, such a response is inadequate. It is to accept that literally no-one ever acts for a j-reason (no-one ever acts for a j-reason that justifies their act). This truth alone is unacceptable, whatever non-literal interpretations one might also give.

We again have a case for taking non-psychologism to be our default position, any other view would have to first unsettle it. However, whilst non-psychologism about j-reasons is quite popular the case with m-reasons comes under rather more attack as I will now consider.

In order to act for the fact *x*, one must believe *x* to obtain. This necessity could be taken to support psychologism. The need for the belief is explained by its identity with the m-reason.[29] However, this support is very weak. The need for belief could be explained in a number of ways. For example, as seems to be the case, the belief may be necessary because it enables one to be motivated. One must believe *x* to obtain in order to act for *x* because believing *x* to obtain enables one to act for *x*. It enables because the agent is aware of *x* through this belief and an agent must be aware of *x* to have it as an m-reason. It would be a mistake to identify enabling conditions with what they enable, so it would be wrong to identify an m-reason with being aware of a fact and hence with having a belief. The role of the belief means that it is possible for the belief to be part of an explanation of the action. But as already noted, it is possible for e-reasons and m-reasons to come apart and we should not let the fact that the belief is an e-reason here make us conclude that it is also an m-reason.

The case with desires is similar though not identical because it is less clear that desires are simply enabling conditions. Rather, those opposed to non-psychologism often take desires to be required for actions on the

Humean grounds that desiring is being driven and acting for a reason is being driven.[30] Smith[31] gives the argument: having an m-reason is, *inter alia*, being motivated; being motivated is, *inter alia*, desiring; desiring is having a desire; therefore, desires are m-reasons. This, however, is not a valid argument. The conclusion one is justified in reaching is that having an m-reason is having a desire. But, unless the having is the same in both these cases it does not follow that the desire is the m-reason. Dancy (2000, 91) shows that it is not the same by pointing out that having an m-reason is being motivated by an m-reason. On the other hand, it is not clear that having a desire is being motivated by a desire. Therefore, the havings are not the same. This is not clear precisely as it is not clear that m-reasons are desires. A desire need not be what motivates it might instead simply be the state of being motivated. This would be sufficient to account for the need for desires in motivated actions and further for the sense in which we can say, as the Humean does, that having an m-reason is desiring.[32] As with beliefs, a desire might explain an action—it might be an e-reason—but it does not follow from this fact that it is also an m-reason.[33] (I will say more on the role of desires in the next chapter.)

That is to say, one cannot disprove non-psychologism by insisting that beliefs and desires are required for motivated actions. Nor can one do so by adopting a Humean stance according to which desire is *the* active state.

However, it is clear that we often speak of people acting because of beliefs of theirs: I went to the hall because I thought it was Tuesday, she was angry as she thought she had been forgotten, etcetera. Furthermore, this way of speaking appears to be forced upon us in those cases where the agent acts due to a mistaken belief. Suppose that Bella was in the supermarket with Tom, she was also following the trail of sugar, and also suddenly stopped to rearrange the shopping in her trolley. We can explain Tom's doing so by referring to the fact that he was making a mess. This was a j-reason for him to stop and he realized as much and was motivated to stop. Bella, on the other hand, was not making a mess but she did form the belief that she was and stop as a result. It appears that we cannot explain her stopping by referring to the fact that she was making a mess because there was no such fact. In this case we explain Bella's act by saying that she stopped because she believed that she was making a mess. Her belief is an e-reason for her act and it looks like it must also have been her m-reason. Furthermore, there appears to be little ground for taking Bella's case to be essentially different from Tom's. Both have beliefs playing similar roles and therefore Tom's belief must in fact of been his m-reason too. I shall refer to this as the *error case* and I take it to be the strongest argument against non-psychologism.

At this point we can emphasize the distinction already noted between a belief and the fact that the belief is held. It might be that it is the fact that the belief was held, not the belief, that provides the explanatory force. Furthermore, psychologism's error case argument assumes that Bella's m-reason is what is referred to when we explain Bella's action with reference to her belief. It assumes that her m-reason is the explanans or e-reason. However, we know that e-reasons and m-reasons can part company. I could explain Bella's act by referring to contractions of her muscles but these were clearly not her m-reasons. In the present case it is clear that the e-reason was not Bella's m-reason. If we had asked Bella what was motivating her Bella would not have cited the fact that she believed she was making a mess, but instead her making a mess. This latter is also what she would have considered in deciding to act.

Psychologism was perhaps encouraged to identify the e-reason with Bella's m-reason on the grounds that the explanation of Bella's behaviour does reveal something about her m-reason. However, it is not necessary for the explanation to do this that the e-reason cited is the m-reason. In this case, the m-reason is revealed by the e-reason because it is what was believed by Bella. That is, the e-reason is the fact that Bella believed she was making a mess, the m-reason is the object of this belief.[34]

Nevertheless, psychologism might still insist that Bella did not make a mess, this fact did not obtain, and therefore it cannot have been what motivated Bella, *pace* my implication to the contrary. The problem as psychologism sees it is that what does not exist cannot motivate.

Lowe (2008, 184–185) has suggested a possible response to this which uses Plantinga's (1978) distinction between states of affairs and facts according to which facts are states of affairs that obtain. We could take it that Bella was motivated by the state of affairs *Bella making a mess*, something which does exist though it is not a fact as it does not obtain. Tom is similarly motivated by the state of affairs *Tom is making a mess*, a state of affairs which actually does obtain and so is a fact. If this view were adopted, non-psychologism about m-reasons would have to be altered to allow that m-reasons can be facts and non-obtaining states of affairs. (Non-psychologism about j-reasons need not alter at all because Bella does not have a j-reason for her action.)

An alternative response, and the one I shall assume, is to deny that Bella had an m-reason at all, but also to deny that this prevented her from being motivated. This response is supported by Bella's willingly citing something non-existent as her m-reason. If we had asked Bella why she was stopping she would have said "I am stopping because I am making a mess". She would have cited something non-existent as her m-reason: she would have cited nothing has her reason.

It is possible to have beliefs concerning something that does not exist. Bella had a belief about her making a mess, something that did not exist. One can take a number of attitudes towards the non-existent, such as when I am afraid of the monster under the bed or when Bella is ashamed of her having made a mess. Therefore, we ought to expect that Bella can be motivated by something that does not exist, or more accurately, be motivated without there being anything that motivates or without having an m-reason. Being motivated is in this case simply another example of adopting an attitude towards a non-existent. A matter of taking something to be a j-reason or to speak in favour of an act and of being moved to act in so taking things. To this end one clearly does not need to have an m-reason to be motivated.[35] (One should not confuse two issues here. Even if there is a metaphysical demand that beliefs about non-existents are constituted by an existent, such as a proposition, it does not follow that the proposition is what they are about or that in this sense they are about anything at all.)

Moreover, we are free to evaluate Bella's behaviour on the grounds of her belief, not because we can assess her m-reason—she did not have one—but because we can consider what would have been the case if her beliefs were true. If Bella's belief that she was making a mess was true, she would have had an m-reason that was a j-reason. Because of this her act appears to be rational even though it is in fact without reason. (I will say more about these issues of rationality in Chap. 6.)

The idea that what does not exist cannot explain is linked with the idea that what does not exist cannot motivate. However, it is clear that this is not a criticism of non-psychologism as I have defended it because the e-reason does exist. The e-reason is the fact that Bella believed she was making a mess. Suggesting that there is no m-reason does not entail that there is a non-existent e-reason, but rather that one cannot explain the case in terms of m-reasons.[36]

It is the case that we sometimes refer to beliefs even in non-error cases: I thought you would be here, John knew the boat was leaking, and so on. Advocates of psychologism might wonder why we would do this if the belief was not the m-reason. Alvarez (2008) offers three explanations, though I suspect there are more. Firstly, one might simply be concerned to bring one's audiences' attention to that belief. (One might be concerned to reveal the psychological mechanisms at play in acting as well as the m-reason.) Secondly, one might want to be withholding one's judgment about whether the belief is true. (I might doubt that the road really is closed, but still take that to be motivating the bus driver.) Thirdly, one might want to reveal the epistemic nature of the agent (as is particularly clear when one uses 'knows' rather than 'believes'). Crucially, however, none of these explanations entail that the belief referred to is the m-reason. So non-psychologism can readily accept that people speak of beliefs in describing actions.

Dancy, has offered us a different sort of example in which it does seem appropriate to speak of belief in citing a j-reason.[37] If Jo has the paranoid belief that kettles are after her, then this might lead her to remove all kettles from her house and avoid tea rooms. However, the fact that she has this belief is a j-reason for her to go to see a psychiatrist because she is clearly unwell. As a j-reason, this must be a fact. So when we say "Jo ought to go to see a psychiatrist because she believes kettles are after her" it is clear that the phrase 'she believes kettles are after her' refers to a fact. Given that this phrase so refers here, we have grounds for believing that our phrasing in reference to Bella—'she believed that she was making a mess'—also refers to a fact. That is, even if such phrases can refer to belief states, it is clear they can also refer to the fact that someone has a belief and we have grounds for thinking that they do so when we use them to rationalize an agent's behaviour.[38]

Jo's example can also be used to ground an attack on psychologism. If Jo was to be confronted about these paranoid beliefs she might be carefully persuaded that she ought to go to see a psychiatrist. If so, she will act for the j-reason cited. She will take the fact that she believes kettles are after her as an m-reason to go to the psychiatrist. This lends further credence to non-psychologism because it is now clear that cases in which one's belief, or rather the fact that one has that belief, is one's m-reason are unusual cases. This undermines psychologism's view that all m-reasons involve beliefs.[39]

There are thus clear reasons to adopt non-psychologism about m-reasons. Further, it is clear that a number of arguments which appear to support psychologism about m-reasons fail to meet their target. Non-psychologism can accommodate the idea that beliefs and desires are e-reasons for an agent's action. Beliefs are necessary conditions for acting for a reason, but they are not themselves m-reasons. Desires are states of being motivated, but they are not themselves m-reasons. Non-psychologism can also account for error cases in which an agent is mistaken about what facts obtain. In such cases an agent simply has no m-reason, though the agent's belief might still be an e-reason for the agent's act.

I thus conclude that j-reasons and m-reasons ought to be taken to be facts. In the next two chapters we will see how this view of reasons fits with the notion of rational action.

Notes

1. Wallace (2003), see also Davis (2003), Smith (1994, 2003), and Mele (2003).
2. Cf. Hurley (2001) who gives an argument that suggests that internalism needn't entail psychologism.
3. It is worth noting that Lowe (2008) uses the term 'externalism' rather differently and in a way more akin to my use of 'non-psychologism' (though Lowe goes on to talk of needs rather than just facts).
4. I do not mean to take a stance on the nature of beliefs and desires by speaking of mental states as opposed to acts or episodes (cf. Steward (1997)).

5. Cf. Alvarez (2010, 3).
6. For arguments to this effect see Strawson (1992).
7. Cf. Williams (1979), Smith (1994), and Korsgaard (1986), and for discussion McNaughton and Rawling (2004).
8. At least of the internalist views that will be amenable to non-psychologism. See Dancy (2000, ch. 7) for a discussion of internalism from the point of view of one who adopts non-psychologism.
9. Bittner (2001) does offer a third alternative by referring to reasons as events and states of affairs, rather than facts. But on the crucial issue of whether one ought to jump because there is a car approaching them or because one believes there is, Bittner, like me, would suggest the former. Bittner does not adopt psychologism about reasons. I will say more about my ontology of facts in Chap. 7.
10. This point is made by many writers, but perhaps especially emphasized by Kim (1998) and Manson (2004) who focus on the first-person perspective.
11. I take the example from Alvarez (2008, 2010, ch. 5.2).
12. These arguments also stand against the minority idea that reasons are propositions. (Cf. Morganti and Tanyi (2017) who explicitly defend this view. Alvarez (2010, ch. 2.2) defends a weakened version of the view because she takes true propositions to be facts. Lowe (1998, 234) criticizes the view Alvarez adopts. The central argument for taking reasons to be propositions is that reasons can play a role in reasoning and it is propositions that play a role in reasoning. But, this argument is no good. Firstly, it confuses what is considered in reasoning (the facts) with the vehicle through which it is considered (the proposition). Secondly, as Lowe (2008, 205–206) emphasizes, reasoning or argument concerns truth preservation: a valid argument preserves truth. Propositions are what can be true or false but it is facts that make them so. Therefore, it is facts that are the aim and value—reasons—in reasoning.
13. Cf. Bratman (1987, 24–27) and Broome (2001).
14. Many details perhaps need to be filled in here to make the CNR plausible, but I set these aside. The point is that one takes Si to be under an obligation to call around because he has this belief. To deny this much would be to remove the criticism of non-psychologism, to accept this much is to accept that the details can be filled in.
15. Cf. Broome (2001). See also Dancy (2000, chs. 2 and 3, 2003).

16. Many of the moves I make here reflect moves made by Dancy (2000). Piller (2003) has criticized Dancy on the grounds that these CNRs merely enable one to criticize someone who breaks them for being inconsistent. Dancy's (2003) response is to say that these CNRs can themselves be grounded, as I have here, and these grounds enable a much stronger criticism of an agent breaking the CNR. Si wouldn't simply be inconsistent if he didn't go to see his mother, he would appear to be doing something in conflict with all she has done for him and the needs a person has when ill.
17. Cf. Dancy (2000, 32) and e.g. Bratman (1987, 23–27), Raz (2005), Wallace (2001), Broome (2005), and Parfit (1997).
18. The bootstrapping criticism could also be applied to this case. The desire to murder is not a reason to murder or to buy a knife.
19. Cf. e.g. Darwall (2003).
20. Cf. Davidson (1973/2001d) and Kim (1988, 392–393). If one adopts a (later) Wittgensteinian view of language we find a similar point being made, but now in the idea that shared understanding must presuppose a shared background of judgements (cf. Wittgenstein (1953/2001, pt. I, sect. 242)). The Wittgensteinian approach makes it clear that this need not lead us to some form of psychologism as the judgments at issue are essentially world involving (cf. Wittgenstein (1953/2001, pt. I, sect. 241)).
21. Further, non-psychologism can readily accept that values can only be derived from values even if these values are all facts. Cf. Scanlon (2014, ch. 2.3–4) who points out we should readily expect the failure of some derivations.
22. Of course there is slightly more to it than this, as from *P* one can always derive *P* or *Q*. Therefore, the evaluative expressions must be occurring in the conclusion in a sufficiently active way.
23. One might argue that derivations concern propositions, not statements. One might also argue that co-referring names will be captured by the same propositional elements and hence try to argue that if the names co-refer there is a sense in which one must be derivable from the other. But, my point is that it is not always clear what propositions are at issue when one is trying to actually perform a deduction, nor whether an act of derivation is logically or merely practically impossible.
24. Though for reasons I will stress in the next chapter I do not take this to be the case, even if reasons are causes, an explanation in terms of reasons is not a causal one.

25. Cf. Cockburn (1997, ch. 5.6).
26. Lowe (2008, ch. 9), see also McCall and Lowe (2005).
27. Cf. Dancy (2003).
28. Cf. Wallace (2003) and Davis (2003).
29. Cf. Alvarez (2008) and Dancy (2000, ch. 6).
30. This can be put a number of different ways: desiring is having a pro attitude and being motivated is having a pro attitude, desiring is active and motivation is active, desiring considers the world the way it could be and so does being motivated, and, desire and motivation have a similar direction of fit (one that is opposed to the direction had by belief, belief being passive and taking the world simply as it is). (Cf. Davidson (2001a), though Davidson is cautious, e.g. Davidson (1978/2001c), Hume (1739–40/1985, bk. II), Dancy (2000, ch. 2), and Smith (1987).)
31. Smith (1994, 116), see also Dancy (2000, 91).
32. Lowe (2008, 186) suggests that desiring can be a mode of awareness of a fact (or need), to this extent it is like a belief and we can see it as a mere enabling condition for acting for that fact much as a belief is.
33. Cf. Alvarez (2010, especially ch. 4).
34. Cf. Alvarez (2008) and Manson (2004).
35. Cf. Alvarez (2008) and Manson (2004).
36. However, Dancy (2000, especially ch. 6.3) does suggest that explanations might not be factive.
37. Dancy (2000, 125), see also Lowe (2008, 181), Alvarez (2010, 48), and perhaps Hurtig (2006, 540).
38. Cf. Alvarez (2009).
39. Cf. Dancy (2000, 125).

References

Alvarez, M. (2008). Reasons and the Ambiguity of 'Belief'. *Philosophical Explorations, 11*, 53–65.

Alvarez, M. (2009). How Many Kinds of Reasons? *Philosophical Explorations, 12*, 181–193.

Alvarez, M. (2010). *Kinds of Reasons: An Essay in the Philosophy of Action*. Oxford: Oxford University Press.

Bittner, R. (2001). *Doing Things for Reasons*. Oxford: Oxford University Press.

Bratman, M. E. (1987). *Intention, Plans, and Practical Reason*. London: Harvard University Press.

Broome, J. (2001). Are Intentions Reasons? And How Should We Cope with Incommensurable Values? In C. Morris & A. Ripstein (Eds.), *Practical Rationality and Preference: Essays for David Gauthier* (pp. 98–120). Cambridge: Cambridge University Press.

Broome, J. (2005) Have We Reason to do as Rationality Requires? A Comment on Raz. *Journal of Ethics and Social Philosophy, 1*(1), (electronic publication).

Cockburn, D. (1997). *Other Times*. Cambridge: Cambridge University Press.

Dancy, J. (2000). *Practical Reality*. Oxford: Oxford University Press.

Dancy, J. (2003). Replies. *Philosophy and Phenomenological Research, 67*, 468–490.

Darwall, S. (2003). Desires, Reasons, and Causes. *Philosophy and Phenomenological Research, 67*, 436–443.

Davidson, D. (1963/2001a). Actions, Reasons, and Causes. In *Essays on Actions and Events* (pp. 3–19). Oxford: Oxford University Press.

Davidson, D. (1973/2001b). Freedom to Act. In *Essays on Actions and Events* (pp. 63–81). Oxford: Oxford University Press.

Davidson, D. (1978/2001c). Intending. In *Essays on Actions and Events* (pp. 83–102). Oxford: Oxford University Press.

Davidson, D. (1973/2001d). Radical Interpretation. In *Inquiries into Truth and Interpretation* (pp. 125–139). Oxford: Oxford University Press.

Davis, W. A. (2003). Psychologism and Humeanism. *Philosophy and Phenomenological Research, 67*, 452–459.

Hume, D. (1739–40/1985). *A Treatise of Human Nature*. London: Penguin.

Hurley, S. (2001). Reason and Motivation: The Wrong Distinction. *Analysis, 61*, 151–155.

Hurtig, K. (2006). Internalism and Accidie. *Philosophical Studies, 129*, 517–543.

Kim, J. (1988). What is 'Naturalized Epistemology?'. *Philosophical Perspectives, 2(Epistemology)*, 381–405.

Kim, J. (1998). Reasons and the First Person. In J. Bransen & S. E. Cuypers (Eds.), *Human Action, Deliberation, and Causation* (pp. 67–88). Philosophical Studies Series 77. London: Kluwer.

Korsgaard, C. M. (1986). Skepticism about Practical Reason. *Journal of Philosophy, 83*, 5–25.

Lowe, E. J. (1998). *The Possibility of Metaphysics*. Oxford: Oxford University Press.

Lowe, E. J. (2008). *Personal Agency*. Oxford: Oxford University Press.

Mackie, J. L. (1977). *Ethics: Inventing Right and Wrong*. London: Pelican/Penguin.

Manson, N. C. (2004). Reason Explanation a First-Order Rationalizing Account. *Philosophical Explorations, 7*, 113–129.
McCall, S., & Lowe, E. J. (2005). Indeterminist Free Will. *Philosophy and Phenomenological Research, 70*, 681–690.
McNaughton, D., & Rawling, P. (2004). Duty, Rationality, and Practical Reasons. In A. Mele & P. Rawling (Eds.), *The Oxford Handbook of Rationality* (pp. 110–131). Oxford: Oxford University Press.
Mele, A. R. (2003). *Motivation and Agency*. Oxford: Oxford University Press.
Morganti, M., & Tanyi, A. (2017). Can Reasons be Propositions? Against Dancy's Attack on Propositionalism. *Theoria, 83*, 185–205.
Parfit, D. (1997). Reasons and Motivation. *Proceedings of the Aristotelian Society, Supplementary, 71*, 99–130.
Piller, C. (2003). Two Accounts of Objective Reasons. *Philosophy and Phenomenological Research, 67*, 444–451.
Plantinga, A. (1978). *The Nature of Necessity*. Oxford: Oxford University Press.
Raz, J. (2005). The Myth of Instrumental Rationality. *Journal of Ethics and Social Philosophy, 1*(1), (electronic publication).
Scanlon, T. (2014). *Being Realistic about Reasons*. Oxford: Oxford University Press.
Searle, J. R. (1964). How to Derive "Ought" From "Is". *The Philosophical Review, 73*, 43–58.
Smith, M. (1987). The Humean Theory of Motivation. *Mind, 96*, 36–61.
Smith, M. (1994). *The Moral Problem*. Oxford: Blackwell.
Smith, M. (2003). Humeanism, Psychologism, and the Normative Story. *Philosophy and Phenomenological Research, 67*, 460–467.
Steward, H. (1997). *The Ontology of Mind: Events, Processes, and States*. Oxford: Oxford University Press.
Strawson, P. F. (1992). Explanation and Causation. In *Analysis and Metaphysics* (pp. 109–131). Oxford: Oxford University Press.
Wallace, R. J. (2001). Normativity, Commitment, and Instrumental Reason. *Philosophers' Imprint, 1*, 1–26.
Wallace, R. J. (2003). Explanation, Deliberation, and Reasons. *Philosophy and Phenomenological Research, 67*, 429–435.
Williams, B. (1979). Internal and External Reasons. In R. Harrison (Ed.), *Rational Action: Studies in Philosophy and Social Science* (pp. 17–28). Cambridge: Cambridge University Press.
Wittgenstein, L. (1953/2001). *Philosophical Investigations* (G. E. M. Anscombe, Trans.). Oxford: Blackwell.

5

Acting for a Reason

In the last chapter I examined the nature of reasons and showed that justificatory (j-) reasons are facts. It was also shown that motivating (m-) reasons are facts and that often one and the same fact both motivates and justifies an action. It is the fact that Mike's meeting starts shortly that justifies his leaving his office. If Mike becomes aware that his meeting starts shortly, then he can become motivated by that fact and act for that reason (if an agent acts for a reason, then that reason is one of the agent's m-reasons).

The present chapter is concerned with the topic of acting for a reason. This takes us beyond the scope of the last chapter because to act for a reason is not simply to behave in a manner that accords with what there is a j-reason to do. If I am being approached by a speeding car I have a good j-reason to jump out of the road. But if I am given an electric shock causing my muscles to spasm in such a way as to propel me out of the road I will not be acting for that reason.

I will not provide a fully worked out metaphysics of what it is to act for a reason. Instead I will bring to the fore a number of important issues and show how these form a coherent view. In doing so I will argue that when an agent acts for a reason it is necessary for that agent to have beliefs and

Olley (F.O.C.H.) Pearson, *Rationality, Time, and Self*,
https://doi.org/10.1007/978-3-319-71973-3_5

desires, and for these to fit together in a specific way. This reflects the fact that when an agent acts for a reason that agent not only has m-reasons but also goals that his or her acts are directed towards. These goals will structure the reasons an agent considers when performing instances of practical reasoning because all of these reasons will be relevant to these goals, however, the goals will not appear as premises or reasons. The conclusion of such a piece of practical reasoning will be an action not a belief about what one ought to do, though of course a consideration of reasons might also lead an agent, in a parallel manner, to form such a judgement.

When we explain an agent's behaviour in terms of the reasons the agent acts for, and perhaps the goals the agent has in acting, this is not a causal explanation. Causal explanations are law like or nomological and explanations in terms of reasons are not. This shows that the causal influences of a reason are not what determine whether or not it is a reason that an agent acts for. However, this leaves open the question of whether or not a reason does happen to cause a particular act.

5.1 Ingredients

Perhaps the best known theory of what it is to act for a reason is that offered by Davidson (1963/2001a). Davidson's characterization has two elements. Firstly, it is suggested that the reason for which an agent performs an action, *x*, is constituted by a pro attitude and a belief. More specifically, a pro attitude towards actions with a specific characteristic and a belief that *x* has that characteristic. Secondly, it is suggested that the reason for which an agent performs an action will be its cause.

Davidson clearly takes reasons, or at least m-reasons, to be psychological states. The arguments of the previous chapter tell us that this is incorrect and to this degree we can dismiss Davidson's view. We can also dismiss a number of other views that take a similar form, such as those offered by Wallace (2003), Davis (2003), Smith (1994), and Mele (2003). However, there are a number of aspects to this Davidsonian view and it is not clear that the arguments of the previous chapter warrant a rejection of all of them. Davidson's view also tells us that when acting for a reason: an agent will have a belief, will have a pro attitude, this belief and pro attitude will stand in a specific relation to one another, and further, this belief and pro attitude will together cause the action. It is possible to accept all of these ideas and to hold non-psychologism about m- and j-reasons. In the following sections I will consider each of these ideas.

5.2 The Need for Beliefs & Desires

We must distinguish the idea that a belief and a pro attitude are necessary for an agent to act for a reason, and the idea that the belief and the pro attitude are reasons, Humeanism.[1] In the previous chapter the latter view was shown to be incorrect but many of the arguments given to support that view can go some way to supporting the idea that beliefs and pro attitudes are necessary in order for an agent to act for a reason.

It is clear that in order to act for a reason an agent must be aware of that reason. If there is a car approaching me then I have a good j-reason to jump out of the road, but if I am unaware of this j-reason I cannot act for it and it cannot be an m-reason for me. I might jump out of the road for some other reason, perhaps I drop some papers onto the pavement. In so acting my action will also be in accord with the j-reason that is the fact that I am being approached by the speeding car, but this is a coincidence. I do not take up this latter j-reason and act for it and it is not an m-reason for me.

Belief is a paradigmatic state of awareness. Often when one is aware of a fact one has a belief about it and vice versa. However, there are also a number of other cognitive—that is representational—states, such as knowing and suspecting. Therefore, one cannot simply move from the fact that an agent must be aware of a reason to act for it to the conclusion that an agent must have a belief when acting for a reason. That said, it is common in the literature to use the term 'belief' as an umbrella term also covering these other cognitive states.[2] This is a practice I adopted in the previous chapter and which I will continue, and with this particular turn of phrase we can conclude that an agent must have a belief in order to act for a reason.

Following this line of argument we can also say something of the nature of the belief at issue: it will be a belief about the agent's m-reason or more specifically a belief with the m-reason as object. In order for the fact that an agent is being approached by a car to be that agent's m-reason to jump out of the road the agent must have a belief with that fact as object. The agent must believe that she or he is being approached by a speeding car.[3]

In the last chapter I adopted the common practice of using 'desire' as an umbrella term which is also supposed to cover states such as wanting and being inclined towards.[4] In this way we can take the notion of desire to coincide with Davidson's notion of a pro attitude.[5]

The idea that a desire is necessary in order for an agent to act for a reason is supported by the fact that if a person does something deliberately then that person must have had some inclination towards doing it. Perhaps I reluctantly wash up my breakfast bowl: I do not like washing up and do not desire (in the narrow sense) to wash up, but I am nonetheless drawn to do so. It would appear quite odd if someone insisted he or she did something deliberately but with no inclination at all to do so. To this end it appears an agent will have a desire (in the umbrella sense) when acting for a reason. As Alvarez (2010a, 119) points out, the state of desiring is a state that a person is in when he or she is motivated.

The distinction between desire as a state of motivation and as an m-reason is one we met in the previous chapter. There I raised an argument presented by Smith (1994, 116) to support Humeanism. The argument runs as follows: having an m-reason is, *inter alia*, being motivated; being motivated is, *inter alia*, desiring; desiring is having a desire; therefore, desires are m-reasons. I responded to this argument with reference to Dancy (2000, 91) who points out that the only relevant conclusion that can be drawn from these premises is that having an m-reason is, *inter alia*, desiring, not, as Smith had intended, that desires are m-reasons. This valid conclusion[6] complements the idea that desires are states of being motivated and entails that desires are necessary in order for an agent to act for a reason. I stand by this response to the argument, but here I want to consider some other issues raised by Smith's discussion.

Smith takes beliefs and desires to be states with a different direction of fit.[7] He suggests that a desire is a state with a world-to-mind direction of fit, whilst a belief is a state with a mind-to-world direction of fit. The idea is that if the world does not match the way you desire the world to be, then you should change the world to fit the desire.[8] Whereas, if what you believe does not match the way the world is, then you should alter your belief. A world-to-mind direction of fit is required in acting for a reason and thus a desire is required in acting for a reason. I think there is something right in this metaphor, but it shouldn't be taken to carry too much weight as it encourages the idea that a desire is faulty when it does not match the world in the same way a belief is when it is false. However, this is surely wrong. Most often we are quite right to desire what is not the case.[9] Further, it implies that having a desire justifies one in changing the world to meet one's desire, but non-psychologism, as defended in the previous chapter, shows this to be wrong.[10]

To further clarify these notions Smith (1994, 115) gives a counterfactual definition of belief and desire. Smith suggests that a belief that *p* and a desire that *p* will have a different counterfactual dependence on a perception that not-*p*. The perception will tend to make the belief go out of existence, but the desire will endure. This point accompanies a dispositional account of desires according to which to desire something is to be disposed to act in certain ways as revealed by such counterfactual dependencies (Smith 1994, 113). However, these ideas should also be taken with caution. It seems quite plausible that an agent may lose a desire upon realizing that what was desired did not obtain. For instance, if I desire to be talking to my sister, but discover I am actually speaking to my mum, then I might well simply alter my desire (it is good to talk to either of them and since my mum is there…).[11] More generally the account is open to the criticisms of any counterfactual account of dispositions. For instance, counterfactuals may fail to hold in finkish or antidote cases where the appropriate circumstances for the disposition to manifest arise, but the disposition is otherwise prevented from manifesting.[12]

Still, I think that there is something right in both of Smith's characterizations. Making these ideas precise would be very difficult, but I take it that we are familiar enough with the notions to make sense of the issue at hand. Desire does appear to be a state of being motivated and to this end it goes hand in hand with a tendency to act and change the world in so acting. Moreover, it differs from belief in just these respects. This is not to say that one cannot be motivated by a reason that one is aware of in having a belief, nor that belief is in this manner impotent. Rather, as desire is the state of being motivated, not what motivates, it is clear that what motivates must be found elsewhere. When one is motivated by a j-reason—when one believes that reason to obtain and hence adopts it as an m-reason—one is thus at once also in a state of desire. The fact that one must be in a state of desire if one is in a state of belief concerning an m-reason that is motivating one—this necessary covariance—does not entail that the desire is identical with the belief or with what it is a belief of (that is, or with the m-reason).

One might be hesitant to accept this identification of desiring with a state of being motivated on the grounds that we on occasion appear to have desires but lack motivation. For example, I might desire something that I know that I cannot obtain like desiring that I did not just drop my bowl on the floor.[13] However, there is a difference between having a desire to *F* and not *F*ing, and having a desire to *F* and not being motivated to *F*. Motivation is such that a motivation to *F* can be manifested in the consideration of the means to or implications of *F*ing. This consideration might in turn mean that one's motivation to *F* is outweighed by other conflicting motivations or, as in the case at hand, is halted at that stage as there simply are no means to *F*ing. An unwillingness to *F* or to consider means or implications of *F*ing is evidence that an agent does not in fact desire to *F* (even if this is inconclusive). This link is what has driven Humeanism and underlies Smith's arguments. I take this link to be real and I have shown that it does not in fact entail Humeanism.

Accepting non-psychologism about j- and m-reasons therefore does not entail that an agent does not need a belief or a desire in order to act for a reason. Rather, there are good grounds for supposing that an agent will always have a belief and a desire when she or he acts for a reason. In the next section I will consider in more detail the role these beliefs and desires play.

5.3 The Role & Interrelations of Beliefs & Desires[14]

A j-reason to do something is a fact that makes that action reasonable.[15] An action can be reasonable in two different ways: intrinsically or instrumentally. For example, I enjoy going for walks and that fact is a j-reason for me to walk. The action of walking is an end in itself. It is intrinsically reasonable. On the other hand, there is nothing intrinsically reasonable about putting my shoes on. But the action of putting my shoes on is a means to going for a walk. The fact that it is a means to walking, which is reasonable, is a j-reason to put on my shoes. Putting on my shoes is instrumentally reasonable because it is a means to a reasonable end.[16]

There must be actions which are intrinsically reasonable if there are actions which are instrumentally reasonable. Instrumental actions are means to ends. However, if there were no intrinsically reasonable actions it would follow that there were no actions that were the fulfilment of ends, and hence that instrumental actions were without ends. The fact that putting on my shoes is a means to walking is a j-reason to put on my shoes only in combination with the fact(s) that makes walking reasonable (in this case this is intrinsic, but of course the chain could be much longer). One does not have a j-reason to do all things which are means to other things. My collecting a saw might be means to my chopping my bed in half, but this does not give me a j-reason to collect a saw because I have no j-reason to chop my bed in half (that is not an end).[17]

I will not delineate exactly what renders an act intrinsically reasonable, but we can find various examples in the literature. Alvarez (2010a, 15, 2010b), perhaps following Aristotle and Aquinas, speaks of actions which involve goods such as pleasure, health, beauty or friendship. In a different manner Lowe speaks of j-reasons not as facts but as needs, such as the need to drink (2008, ch. 10). Lowe's suggestion could be transposed as the suggestion that acts which are the fulfilment of needs are intrinsically reasonable. More generally Parfit and others have defended a substantive view of rationality according to which all rational agents must care about particular things such as their own health.[18] For example, people who will gain nothing from dying appear to be irrational if they do not recognize any value in taking a medicine that they have every opportunity to take to prevent themselves from dying. Again, this view might be transposed as the suggestion that acts directly responsive to these substantive demands are intrinsically reasonable. I will follow Alvarez and to this end an act that gives pleasure,[19] promotes health, or is an act of friendship or beauty, is an intrinsically reasonable act and the fact that the act possesses these qualities is a j-reason to perform that act. Some instrumental actions will be actions that are means to actions which are themselves intrinsically reasonable and the j-reason to perform these acts is the fact that they are such means. Other instrumental actions will be means to instrumental actions of the first sort and so on for further instrumental actions that are means to instrumental actions of the second sort etcetera. But the adoption of a different substantive view of rationality would do little to alter my point.

Very often there will be more than one j-reason to perform an act. Alvarez (2010a, 126–128) is careful to point out that this can happen in two different ways. Firstly, the j-reasons might be independent of one another. For example, my going to a café will enable me to have a coffee [pleasure] and will enable me to see a friend. Either of these facts without the other is a reason for me to go to the café. They each independently pick out something reasonable about my going there. Secondly, we have cases where the j-reasons are not independent. These are cases in which a number of facts together make an action reasonable though independently they would fail to do so. One example of this has already been indicated: instrumental reasons are only j-reasons in combination with intrinsic reasons. The fact that putting on my shoes is a means to walking is only a reason for me to put my shoes on if I also have a reason to go walking. A second example of this concerns facts which support an instrumental or intrinsic reason. For example, putting on my shoes is a means to walking because of various other facts, such as the fact that my feet are delicate. However, the fact that my feet are delicate is not a reason to put my shoes on independently of other facts which along with it support the instrumental reason. For example, the fact that my feet are delicate combines with the fact that walking involves being on one's feet a lot to support the fact that putting on my shoes is a means to walking. We might refer to these as *supporting* reasons.[20]

When people act for a reason they are motivated to act for that reason. We know from the preceding discussion that being motivated to act for a reason involves having a desire and to this degree being motivated to act for a reason implies desiring for a reason. In fact, we can distinguish two varieties of desires: those which are had for reasons and those which simply assail us. I shall refer to the former as *rational desires* and the latter as *non-rational desires*.[21]

It is clear that some of our desires simply assail us. For example, I might suddenly feel the urge to scream or become thirsty. These are non-rational desires (which is not to say they are irrational). However, other desires we have clearly are affected by reasons that we consider. For example, if I wish to become faster at sprinting and I learn that doing squats is a means to becoming faster at sprinting, then this may well lead me to desire to do squats. If that is the case, then this desire to do squats will have arisen because of the reason that doing squats is a means to sprinting. I will have formed this desire for a reason. This is not to say that peoples' rational desires are directly subject to their choices or that we can choose what to desire. It is simply to say that the reasons we consider can give rise to desires which are responsive to them.[22]

When an agent is motivated to perform an action that agent will have a desire and the object of that desire will be precisely to perform that action.[23] Here we must again be mindful of the act/object distinction. When an agent is motivated the agent will be in a state of desire (act), there will therefore also be something that the agent desires (object). This object will be *to perform the action*. Following Alvarez, I will refer to such a desired object as a *goal*. It is the goal with which an agent acts.[24] It is possible that an agent will also have a different goal in acting. For example, when the agent's act is instrumental it will be a means and the agent will desire that means and the end that it is a means towards. However, we may not act to fulfil every desire we have and to this end not all objects of desire are adopted as goals.[25]

Let us consider the case of walking again. Walking is an action that possesses an intrinsic good. It is something that gives me pleasure. The fact that walking gives me pleasure is therefore a reason for me to walk. If this reason motivates me to walk I will desire to walk and my goal in walking will simply be to walk. On the other hand, my putting on my shoes is an action that does not possess an intrinsic good, but it does possess an instrumental good because it is a means to walking. The fact that putting on my shoes possesses this instrumental good is a reason for me to put on my shoes. If I act for this reason then I will be motivated to put on my shoes. I will have both the goals of putting on my shoes and of walking.

Alvarez (2010a, ch. 4.1.1) argues that goals should not be conceived of as reasons because goals are not facts that make an action good. Going walking is not a reason to go walking. Rather, the fact that going walking will give you pleasure is. Similarly, walking is not a reason to put on one's shoes, though the fact that putting on one's shoes is a means to walking is a reason to put on one's shoes. There are thus two distinct elements in acting for a reason: the reasons and the goals. We might say we are walking simply for walking's sake, but it is plausible that this either means that we are walking because we enjoy walking, find it relaxing etcetera, and are not doing it as a means to anything else, or we mean we are simply walking without a reason at all.

When a person is assailed by a non-rational desire and simply acts in response that person is not acting for a reason. This is clear when the act is immediate. For instance, I may simply drink when thirst strikes if I have a drink to hand. We saw in the last chapter that desires were not reasons for acting and so it is clear that in such a case an agent would not be acting for a reason. Nevertheless, people must drink in order to survive. Agents might, upon feeling thirsty, take themselves to require water. They may realize that drinking will be good for their health and hence that they have a reason to drink. In such a case an agent can of course drink for this reason. (It is also possible to explain an agent's action with reference to a non-rational desire that simply assails and compels the agent. A non-rational desire can be an explanatory reason, or e-reason. This doesn't mean it is an m- or j-reason. Moreover, the more accurate thing to say might be that the fact that the agent has that desire is the e-reason.)

There is a clear relationship between our m-reasons and our goals, and hence between the beliefs and desires agents must have when they act for a reason. When an agent performs an action the agent must have that action as a goal. The agent must desire to perform that action and believe that that action possesses either an instrumental or intrinsic good. The agent must be aware of the reason and motivated to perform that act. To this end we can agree with Davidson that when an agent acts for a reason the agent will have a belief and a desire and these will stand in a particular relationship. However, the form of this relationship is quite different from that suggested by Davidson.

5.4 Practical Reasoning[26]

One can reason about whether to pursue a goal, or about how to pursue a goal. However, not every instance of acting for a reason will involve practical reasoning. For example, if I enjoy singing I might simply sing for that reason and this need not require me to do any reasoning. When agents act for an instrumental good they will usually have engaged in reasoning to the extent that they recognize an action to be a means to an end and connect the instrumental and intrinsic reasons together.

Practical reasoning is commonly taken to have a particular form: *b* desires *F*, *G* is a means to *F*, therefore, *b* ought to *G*.[27] (This being a piece of practical reasoning that leads *b* to do *G*.) This view of practical reasoning is open to numerous criticisms. Firstly, the form of the syllogism is too narrow as it does not allow for other desires an agent might have and how these might interact.[28] For example, if *b* also wanted to *H*, and wants to *H* more than he wants to *F*, and the only means to *H* is to not *G*, then one might expect that *b* shouldn't *G*. Moreover, this form, even if it were somehow expanded so as to accommodate other desires, would be subject to a criticism stemming from the bootstrapping argument considered in the last chapter and more generally from non-psychologism about reasons. An agent does not have a reason to *F* or do what subserves *F*ing just because they have a desire to *F*. Desires are not reasons.[29] In short, the conclusion of such a syllogism does not follow from its premises.

The premises of a piece of practical reasoning ought to be reasons to perform an action or rather beliefs about such reasons.[30] If we consider that the conclusion of a piece of practical reasoning is that one ought to act[31]—for example, *b* should *G*—then the premises of good practical reasoning will have to lead to this conclusion. But, considerations that make an act good just are reasons to perform that act, therefore the premises are or express reasons.

Practical reasoning will also involve having a desire just as we saw that acting for a reason did. The desire will again be a state of motivation. It will give rise to the practical reasoning. Moreover, the object of that desire—the goal—will delimit the practical reasoning. If I desire to go for a walk and so have a goal of going for a walk then this will specify the sort of reasoning that I will engage in and the reasons that will appear as premises in that reasoning. As already noted, if I desire to go for a walk this might be simply for the reason that walking is pleasurable for me and I might simply walk without needing to perform any practical reasoning if I do not need to take any means. In this case my reason for walking will be the fact that I find walking pleasurable and this reason will also have given rise to a desire to walk so I will have a goal of walking. If I do need to take means to go for a walk then my goal of walking will lead me to reason in a particular way. It will lead me to consider what the means to walking are. My goal of walking will not appear as a premise in my reasoning. Walking, or to walk, is not a premise leading to the conclusion that one ought to put on shoes and nor is the desire itself such a premise. But, given this goal only particular patterns of reasoning and only the consideration of particular reasons will be initiated or appropriate. These reasons will concern means to walking or the value of walking. Thus, when I am motivated by the fact that walking is pleasurable I will have a goal of walking and if I need to take means to walking then I will perform reasoning about what the means to walking are. I might for instance reason as follows: walking gives me pleasure; walking involves being on my feet a lot; my feet are delicate so being on my feet a lot requires the wearing of shoes; I have no shoes on, so putting on shoes is a means to going for a walk; so I ought to put on some shoes. The desire to walk and the goal of walking do not appear in this reasoning, but they clearly play a role in motivating it and the reasoning/reasons are clearly related to the goal of walking. They are fitted to it.

If one were to reason about which ends to adopt as goals or which objects of desire to pursue in an instance of action, as opposed to how to pursue these goals, then matters will be a little different. Now one might consider the intrinsic or instrumental value in performing a desired action and compare this to the intrinsic or instrumental value in performing a different desired action. One might also consider the means required to perform each of these actions. All of these factors will be premises in one's reasoning and potential reasons for their action of adopting one goal rather than another. In this one will again be motivated and one will have a goal. One's goal might simply be to reach a decision or it might also involve the objects of the desires under consideration. Either way, it is not required that any of these goals or the desires they are the objects of appear as premises in this reasoning. But, the reasons I do consider—the premises of my reasoning—will all be related to one or more of these goals.

As Alvarez (2010b), Anscombe (1957/2000), and others have said, desires or the objects of these desires do not appear in practical reasoning, contra the traditional view. But we have not moved too far away from this view because the premises that do appear will concern these desires. They will be facts concerning the value of goals or the means to achieving those goals.

Broome takes practical reasoning to have a different general form: *b* intends to *F*, *b* believes *G* is a means to *F*, therefore, *b* intends to *G*.[32] In this Broome supposes that intentions and beliefs are both states with propositional content, and this content will form a general pattern: *b* will *F*, *G* is a means to *F*, therefore, *b* will *G*. The preceding discussion concerning taking desires as premises applies equally to taking intentions as premises. Neither are fit for the job. However, if we instead take the premise to be that *b* will *F* matters are different. Nevertheless, it remains the case that the fact that *b* will *F* is not a reason for *b* to *G*. If *b* has no reason to *F* then *b* may have no reason to *G* either. Our lesson concerning CNRs applies here. Even if there is something at fault in *F*ing and not *G*ing it doesn't follow that *F*ing is a reason to *G*. Rather there may be a reason to avoid this combination and not *F*ing is a way of doing that.

A different concern with the traditional understanding of practical reasoning is whether it is in fact at all practical. The worry is that one could conclude that one should *G* without thereby being motivated to *G* and without *G*ing.[33] Practical reasoning is supposed to be reasoning bound up with actually acting not to be merely reasoning about actions. If an agent's reasoning never influenced that agent's actions then it would not be practical reasoning. Reasoning to predict one's own behaviour is not practical reasoning because predicting one's own actions does not have the right connection with acting. Similarly, one might reason to construct a moral theory, but if one does not adopt this theory then that reasoning will be purely theoretical and not practical. Thus one is driven to the view that the conclusion of a piece of practical reasoning is an action.[34] However, the view that the conclusion of a piece of practical reasoning is an action has met criticism.

Broome (2001, 2002) expresses the following idea: reasoning is essentially a mental activity and therefore the conclusion of a piece of practical reasoning must be something mental, such as a mental state; actions are rarely something mental[35] and therefore they cannot be the conclusion of a piece of practical reasoning. This, however, is clearly just question begging. It assumes from the start that actions cannot be conclusions. But this does relate to a more general worry: what is it about practical reasoning so described (with action as conclusion) that makes it an example of reasoning?[36]

It is clear that an action cannot be deduced from premises as it is of the wrong ontological kind, therefore if one must be able to deduce a conclusion from premises to have an instance of reasoning then practical reasoning cannot conclude in an action. But it is clear that not all reasoning does involve deduction: one can reason inductively. Moreover, taking theoretical reasoning as a paradigm of reasoning, theoretical reasoning is an activity. It is generally assumed that logical relations such as entailment hold between propositions and these are distinct from the beliefs, sentences, statements and utterances that can express them. When one concludes a piece of theoretical reasoning one forms a belief which is something of the wrong ontological form to be deduced. In short, theoretical reasoning, as embodied in beliefs, does not strictly have a logical form. Rather, reasoning is responsive to broadly logical form because the propositions expressed by the premise beliefs, if true, will (ideally) be evidence for the truth of the proposition expressed by the concluding belief. This radically closes the distance between practical and theoretical reasoning. An action that is the conclusion of a piece of practical reasoning will not have a propositional content and is in this way unlike a belief which concludes some theoretical reasoning. But, the action which is the conclusion of a piece of practical reasoning will be such that if the propositions expressed by the premise beliefs are true then (ideally) the action will be rational. Neither theoretical nor practical reasoning have a logical form as they are composed of entities of the wrong ontological kind. Rather, they reflect a broadly logical form. In theoretical reasoning this is achieved through beliefs having a propositional content but we needn't assume that this is the only way to reflect broadly logical form. The fact that an action lacks a propositional content does not stop it from being sensitive to the broadly logical relations that hold between propositions expressed by premise beliefs. In having this sensitivity a concluding action is similar to a concluding belief. For example, in reasoning to reach the conclusion of acting to put on my shoes I will consider the fact that walking gives me pleasure, that walking involves being on my feet, that being on my feet is uncomfortable without having shoes on, etcetera. If we say that the propositions capturing these facts are the contents of the beliefs I have in reasoning it is clear that these propositions are related in ways determined by their broadly logical form and it is only because they are so related that we act as we do in conclusion and that this concluding act is a reasonable one (or rather, it is because the facts are so related that the propositions that express them have the logical relations they do).

A different concern with the idea that practical reasoning concludes in action arises from the appearance that one can perform practical reasoning without acting. This might happen if an agent reasons about what to do whilst unknowingly paralysed.[37] However, it is not clear that this undermines the idea that an action can be the conclusion of a piece of practical reasoning. We can distinguish a number of cases. Firstly an agent might reason about what she or he ought to do at some time in the future. For example, going to the meeting will give me pleasure, in order to go to the meeting I must leave my office at 2pm, therefore … . We should not expect this reasoning, if performed at 1pm, to result in my immediately standing up to leave my office because that action is not required for some time. However, at this stage the action takes the form of my efforts to determine subsequent reasoning and actions so that they all enable me to leave my office at 2pm.[38] This gives the action a slightly different form from that of simply leaving the office, but in cases where no time delay is necessary there is no reason to suppose that action cannot be manifested in this latter more direct way.

A second case in which a piece of practical reasoning will not result in an action is a case in which the premises promote inaction. For example, if it appears best to wait a little longer to see if your parcel will arrive. We are here dealing with the case of omissions.[39] The recognition of omissions as of normative significance is growing. Omissions are increasingly recognized as having a relation to reasons, intentions, and the like. Although this is usually in respect of an omission going against the reasons an individual might have to act this also opens up the possibility to see an omission as in fact responsive to reasons. As such, it appears that we have some scope for recognizing an omission as a conclusion of a piece of practical reasoning in the relevant cases. Recognizing an omission as a conclusion is not of course recognizing an action as a conclusion, but it is also clearly not in conflict with allowing that actions can be the conclusions of instances of practical reasoning.

A third example in which practical reasoning may not result in an action is given by the case mentioned above in which an agent is, unbeknownst to herself, unable to act. Any practical reasoning that such an agent could perform apparently could not conclude in her acting. However, we must only say this if we think that all acting must involve movements of the body but it isn't clear it must. We can suppose that the agent begins to act in coordinating her attention etcetera, but she is prevented from carrying such action through to certain bodily movements.

Thus I find no good reason to think that an action cannot be the conclusion of a piece of practical reasoning, rather recognizing conclusions are actions respects the practicality of the reasoning. Practical reasoning will be guided by a goal which unites the premises of the reasoning but which is not itself one of those premises. The premise beliefs will all concern the goal insofar as they will concern facts about the reasonableness of the goal or the means to reaching that goal. The conclusion of this reasoning will be an action (or omission) that embodies (at least in part) one of these means.

None of this is to deny that the consideration of reasons can give rise to certain judgements like the judgment that one ought to *G*. But, this judgment needn't come between the consideration of the reasons and the action. Even if it is a fact that one ought to *G* one needn't be responsive to this fact in order for one to *G*. Recognizing that *G*ing is intrinsically reasonable—for example, is good for one's health—is already sufficient to motivate one to *G* (if it wasn't then it is doubtful it could be sufficient to motivate the judgment either).

If a consideration of reasons gives rise to both a judgment and an action then it is best to think of these as arising in parallel. Similarly, although desires can be responsive to reasons this doesn't mean that they ought to be recognized as *the* conclusion of a piece of practical reasoning, even if one supposes that one must desire if one acts for a reason. Desire is not that which motivates. It is the state of being motivated. As such, it needn't be a stage between the reasons one considers and one's act. It is best considered as parallel to one's act or as involved in it. Both are responsive to reasons. This does not mean that one can choose to desire even if choosing to act goes hand in hand with forming a desire. (Choice might affect the onslaught of the desire without the desire being chosen. There is more to being chosen than being the result of a choice. Moreover, it is appropriate to think that some of the pro attitudes encompassed by our thin umbrella notion of desire will be receptive to our choices.)

5.5 Reasons as Causes

In this section I will consider Davidson's suggestion that a reason an agent acts for (an agent's m-reason) is necessarily a cause of the action done for that reason. I am concerned to argue that recognizing a reason as a reason that an agent acted for doesn't involve recognizing that reason as the cause of the agent's action. In arguing for this I will leave open the question of whether or not the reason does in fact happen to be a cause of the action.

Davidson (1963/2001a) believed that it was necessary that a reason was a cause of an action done for that reason. Davidson's concern was that a person might be aware of a number of different reasons for an action and perform that action, but only perform the action for a subset of those reasons. For example, I might be aware of a number of reasons to cross the road: my crossing the road will enable me to get away from my associate who is irritating me and it will enable me to help a person who has just fallen and is clearly suffering. But I might insist that the reason that I do actually cross the road for is that it will enable me to help not that it would enable me to avoid my associate. We must be able to distinguish between the reasons we act for and the ones we are merely aware of. To this end Davidson suggests that the reasons we act for must be those which cause our action. The reasons we do not act for do not cause our action. If Davidson was right this would imply that recognizing a reason as a reason an agent acted for would involve recognizing that reason as a cause of that act as this is all there is to stand apart those reasons acted for and those reasons merely had in mind. (Davidson was thinking of reasons as belief/desire complexes rather than as facts as I have argued that we ought to, but this makes little difference at present.)

Davidson does not try to argue against other accounts of the connection between a reason and an act performed for that reason, and therefore his argument only has force so long as we take a causal account to be the only possible account.[40] However, Lowe (2008, 183 and 202) offers a rather different account: the reasons that an agent acts for are those reasons that the agent chose to act for or chose in light of. Further, to make sense of this it is not necessary to suppose that the choice itself must have been caused by the reasons. Choice is already a familiar notion which is taken to be responsive to reasons but not caused by them. This is shown by the phenomenology of choosing to act in light of a reason. One does not feel caused to act by the reason one acts in light of. On the contrary, it appears as if the reason is impotent without one's act of choosing. Moreover, one is struck by the fact that taking one's choice to be the causal result of a reason conflicts with the very idea of making a choice.[41] Nevertheless, one's choices are not mere chance events. They are responsive to reasons we consider.

We can discover which reasons an agent chose to act in light of in a number of ways. Very often it is sufficient to simply ask the agent. An agent might try to deceive you or himself but that does not show that there is not a fact of the matter or that we must resort to a consideration of causes to settle the issue. Rather, it is plausible that people can deceive themselves and others about choices they make. Further, we might uncover such hidden choices through considering the reasons an agent is aware of and the agent's character which is given by the agent's past choices.[42] None of this entails that we are thinking of the choices as caused by the reasons.

The notion of choice can be fitted into the account given above by taking the choice to be an action of the agent. The choice can then be seen as the conclusion of a piece of practical reasoning. This does not entail that the conclusion of a piece of practical reasoning cannot be an action like putting on one's shoes because such an action can be recognized as beginning in a choice.[43] There is little value in supposing that an agent must perform an act of choice before the agent can perform a chosen action: it is hard to conceive what such a pure act of choice could be.[44] When a person simply decides to walk and need take no means to walking then this decision is manifested in the person's walking. There does not need to be a gap between the decision and the act itself. When means are necessary an agent's decision to walk will be manifested by actions to uncover and pursue means to walking, and in turn with walking.

An alternative ground for taking the reasons an agent acts for to be causes of that agent's action might occur if the explanation of an agent's action in terms of the agent's reasons for acting was a causal explanation. However, in considering this we must be mindful of a distinction drawn by McGinn (1979) between giving a causal explanation and giving an explanation in which the explanans happens to be a cause of the explanandum. It is possible to explain one thing, *b*, in terms of another, *c*, without *c* being the cause of *b*. In logic and mathematics we do this all the time. If this is done we will have an explanation of *b* but not a causal explanation. An explanans does not need to be a cause. Of course, one can provide causal explanations so an explanans can be a cause. Nonetheless, it is also possible that we explain *b* with reference to *c*, and that *c* does in fact cause *b*, but that the explanation we give is not a causal explanation: the explanation does not assume or utilize that causal connection. So, should we think that explanations in terms of reasons—reasons explanations—are causal explanations?

One argument for thinking that reasons explanations are not causal explanations arises from considering the possibility of deviant causal chains. After giving his account of acting for a reason Davidson later noted that it suffers a difficulty in the form of deviant causal chains. Davidson (1973/2001c) gives an example in which a climber is holding her partner on a rope and the weight of her partner is threatening to pull her off her ledge. The climber might recognize that ridding herself of the weight of her partner would potentially be good for her health and that she could do so by letting go of the rope. This awareness might unsettle the climber causing her to shudder as she becomes horrified of the thought of willingly dropping her partner. In shuddering the climber might in fact lose hold of the rope and drop her partner. If this were to happen it would be clear that the climber did not drop her partner for the reasons she was aware of although these reasons did cause her to let go. Davidson's response to this problem was to suggest that the reasons must cause the action in a specific way in order for the action to be performed for those reasons. It follows that if the causal account is correct then recognizing a reason as the reason an agent acted for involves recognizing that reason as a very particular type of cause of the agent's action. However, it has remained quite elusive just what particular form such causation ought to take.[45] This makes it implausible that reasons explanations are causal. If reasons explanations were causal, then in order to give a reason explanation one would have to determine that a reason was a reason acted for and therefore determine that that reason played a very specific but unknown causal role. The ease with which we uncover reasons explanations and the apparent impossibility of uncovering the relevant causal relation suggests that the former does not rely on the latter and reasons explanations are not causal explanations.

Causal explanations are not simply explanations that explicitly refer to the notion of a cause and so it can be difficult to distinguish causal and non-causal explanations. One way of doing so is to note that causes are lawlike and so a causal explanation ought to be an explanation that falls under a law.[46] But, considering that causal explanations are lawlike gives further grounds for denying that reasons explanations are causal explanations.

Davidson's view was that causal relations can always be subsumed under laws.[47] But, Davidson also believed that reasons were belief/desire complexes and that there were no laws linking these with actions.[48] It is important to note that these views do not lead Davidson into contradiction when he also says that reasons are causes. Rather, Davidson avoids contradiction by suggesting that whether two events are lawlike related depends on how they are described. So, two events might be described in such a way that they are not lawlike related even though those two events are related as cause and effect so long as there is some other way that they can be described such that they are lawlike related. So, as long as the event that is the reason and the event that is the action can both be re-described in such a way that they are lawlike connected the former can be a cause of the latter despite reasons not being lawlike related to actions.

However, although Davidson in this way avoids contradicting himself problems linger for the idea that reasons explanations are causal explanations. Crucially, if reasons and actions are not lawlike related, then we cannot be presupposing a lawlike connection between reasons and actions performed for those reasons when giving reasons explanations. We are left with grounds for denying reasons explanations are causal explanations (even if reasons happen to be causes of actions performed for those reasons).[49]

Not taking reasons to be belief/desire complexes my issue appears to be fundamentally different from Davidson's. However, it has been argued that explanations in terms of reasons are elliptical for explanations citing beliefs and desires because, as already noted, an agent must be aware of any reason he or she acts for.[50] I am not sure that this is right but I accept that beliefs and desires will play a role in acting for a reason and so will consider explanations in terms of beliefs and desires (even though the reasons are the objects of these beliefs).

It is clear that we do not take there to be a lawlike relation between beliefs/desires and an action when we accept an explanation of that action in terms of those beliefs/desires or the reasons the action was performed for.[51] This stems from the idea that an agent chooses to perform that act in light of those reasons and the agent could have chosen to do otherwise. Frankfurt (1969) has famously argued that an agent needn't be able to do otherwise in order to be free or to choose to perform a specific action. But Frankfurt's argument is questionable because it relies on assuming that someone can control how an agent chooses or acts, and it is not clear that this makes sense. It is not clear that what would result from such control could be an action or choice.[52] I will not try to settle this issue here. The point I wish to draw is more general. This very discussion of the possibility of doing otherwise shows that the thought that an agent cannot do otherwise is not straightforwardly derived from the thought that that agent acts for a reason given through—or in—his or her beliefs/desires. It follows that any explanation of an agent's action in terms of the reasons the agent acts for cannot be a causal explanation.[53] The force of such an explanation cannot be derived from a lawlike or causal relation between the reason and the action because the explanation can stand in the face of doubts about any such relation.

Manson (2004) proposes that we needn't take reasons explanations to be grounded in generalizations or lawlikeness at all. An explanation here is simply something that can lead one to understand why an agent performed an action. Because we are all agents and familiar with acting for reasons when we are informed of the reasons an agent acts for we can put ourselves in that agent's shoes and thereby come to comprehend that agent's action. This distinction bears a resemblance to the distinction between the theory-theory and the simulation-theory approaches to interpersonal understanding.[54] On the former view we must construct a theory of how other people work which involves generalizations about relations between beliefs/desires and actions. On the latter view this is not necessary. Rather, because we are agents we can imagine ourselves in another agent's circumstances when we know what that agent believes/desires and we can effectively run a simulation and as a result be led to expect certain actions. In this latter case there is no need to assume lawlike connections hold between belief/desires and actions or to suppose a causal relation holds. Nevertheless, one is brought to understand why an agent performed an action. I know how I would act in John's circumstances and so understand his so acting.

Recognizing a reason as a reason an agent acts for does not require one to uncover a special causal relation between that reason and that action, nor even to presume that such a connection could be uncovered. Though, this is not to assert that no causal connection will happen to hold between a reason and an action done for that reason.

5.6 Summary

I agree with Davidson that when an agent acts for a reason that agent will have a belief and a desire. These states will not fit together as Davidson suggested though. The belief will be about the reason. The desire will be a desire to perform that action.

Recognizing that an agent performs an action for a reason does not involve recognizing that reason to be a cause of that action. Reasons explanations are not causal explanations. Though, this leaves open whether or not the reason is in fact a cause of the action.

Notes

1. Hume (1739–40/1985) appears to say that belief alone (as a mere concern for matters of fact) cannot lead to action, rather we also require a desire (a passion). Whether or not this is a correct interpretation of Hume's position the view that an agent's motivational reasons are a combination of beliefs and desires has become known as 'Humeanism'. Cf. Dancy (2000, 10), Smith (2003), and Cullity and Gaut (1997).
2. Cf. Alvarez (2010a, 10 and 126).
3. Below I will add to this. We may also need a belief such as that stepping out of the road will involve avoiding a speeding car and so benefit my health.
4. Cf. Alvarez (2009a, 55).
5. Cf. Davidson (1963/2001a, 4). Schueler (2003, ch.2) warns against conflating the two because Davidson's notion of a pro attitude includes "*moral views*" (Davidson 1963/2001a, 4) and to this degree also appears to include beliefs such as 'it is wrong to *G*'. However, I shall not dwell on this here. My concern is that the objects of desire are not reasons to act, but are perhaps goals one has in acting. It appears that Davidson takes his pro-attitudes to give the ends of an act, as he talks of having a pro attitude *towards* something. I take this to be enough to put pro attitudes in line with desires. (Though I do want to remain open on whether there are moral facts, such as that one ought to *G*, if there are such facts an agent can be aware of them and act for them, but in this case I would take them to be the objects of beliefs, not of desires or pro-attitudes.)
6. Cf. Alvarez (2010a, 121).
7. Cf. Smith (1987, 1994) and Platt (1979, 256–257) who Smith attributes the idea to and who in turn cites Anscombe (1957/2000).
8. This does assume that desires are propositional attitudes, that is, the content of a desire is a proposition or is captured buy one. This does seem to be a popular view (e.g. Smith (1994) and Broome (2002)). However, Alvarez has argued that this is incorrect. The content of a desire can always be captured by an infinitival clause e.g "I want to *F*" or "I want *b* to *F*", but it is awkward if not misleading to try to capture it with a that clause "I want that *b F*" (Alvarez (2010a, ch. 3 and 4), see also Ben-Yami (1997) and Thagard (2006)). Thagard (2006) also argues that desires are not propositional attitudes on the grounds of evidence from comparative psychology and neuroscience.

9. Cf. Alvarez (2010a, 69).
10. Cf. Alvarez (2010a, 69).
11. Cf. Alvarez (2010a, 69) and Schueler (2003, 34).
12. Cf. Martin (1994) and Bird (1998).
13. Cf. Schueler (2003, 34).
14. I owe much of the account of this section to Alvarez, especially Alvarez (2010a).
15. Cf. Alvarez (2010a, 12).
16. The distinction between intrinsic and instrumental actions is not one original to me (e.g. Alvarez (2010a, ch. 4.2.2)), but instrumental actions have dominated some discussions of rational actions (e.g. Walton (1967) and Beanblossom (1971)).
17. The thought that each action is merely a means to a further action which is itself merely a means to a further one, and so on for infinity, is clearly unacceptable. One would have to see that which an action was a means to as of value to perceive that action itself as of value, and hence, one would have to be able to perceive an infinity of actions as of value in order to perceive any one action as of value and this is obviously not the case. Moreover, this strikes one as an infinite chain of dependence and such chains are highly dubious (there possibility is generally doubted, e.g. Leibniz (1974), Kant (1787/1996, B436-437), Cameron (2008), contra Brown and Ladyman (2009)).
18. Cf. Parfit (2001, 1987, sect. 46) and Dancy (2000, 17).
19. The action is reasonable because of the fact that the action gives pleasure, this is quite different from saying that the action is reasonable because of a desire to perform it, we have not fallen into psychologism about reasons.
20. This needn't imply that there are always innumerable reasons for each act or if it does this is not a problem. Our concern is with the reasons the agent acts for and these will be reasons that the agent is aware of. Although we could introduce a great number of relevant facts in such a case, we needn't become bogged down by them, just as the agent needn't be.
21. I take the terminology from Alvarez (2010a, ch. 3.3), but the distinction can be found in forms in a number of writers, e.g. Dancy (2000, ch. 4), and Nagel (1978, ch. 5).
22. This differs from Nagel (1978, ch. 5), who speaks of choice in regard to his similar distinction between motivated and motivating desires.

23. One might dispute this on the grounds that the objects of such desires ought to be taken to be general so that we should perhaps say that the desire is to perform an act of a certain type, not that particular token. But, this requires little change. Cf. Lowe (2008, 207), Davidson (1963/2001a), and in a related way, Hornsby (1980, ch. 1) in her suggestion that what we do is always something general.
24. Cf. Alvarez (2010a, ch. 4.1.1). This relates to the notion of intention adopted by Broome, e.g. (2009) when he speaks of our intention in acting or intention to act.
25. One might, e.g., decide to pursue a different desire instead. Cf. Alvarez (2010a, 96).
26. I am heavily influenced by Alvarez (2010b) throughout this section.
27. E.g. McGinn (1979) and Schuler (2003, ch. 4).
28. E.g. McGinn (1979) and Schuler (2003, ch. 4).
29. Cf. Schuler (2003, 92).
30. Actually, because one can have reasons both to perform or not perform an action, the premises of a piece of practical reasoning might include both reasons *pro* and *contra* the action.
31. Or simply 'the action one ought to perform' if we adopt the view below that the conclusion of a piece of practical reasoning is an action.
32. Cf. Broome (2001, 2002), and relatedly Rundle (1997, 192).
33. Cf. Rundle (1997, ch. 7).
34. Cf. Alvarez (2010b), Aristotle (384-22BC/2006, ch. 7, 384-22BC/2001, bk. VI), and Anscombe (1957/2000).
35. Of course one might take such mental actions as the formation of a decision or the initiation of the will as actions which are somehow mental (cf. Lowe (2008, ch. 8.4)).
36. Cf. Audi (1989, ch. 1.IV).
37. Cf. Broome (2002), Alvarez (2010b), and Audi (1989, 18).
38. Cf. Korsgaard (2009).
39. Cf. Williams (1995).
40. Cf. Schuler (2003, ch. 3). Schuler himself speaks of an agent's character suggesting that the reasons an agent acts for are distinguished from those he merely considers by being suited to his character.
41. Cf. Lowe (2008, ch. 9 and 10).
42. Cf. Schuler (2003, ch. 3).
43. Cf. Korsgarrd's (2008) identifying the intention/judgment with the action and relatedly Korsgaard (2009, ftnt. 3).

44. There is here perhaps an issue of what an action is. Should it involve the world as putting on one's shoes does, or should it stop at the boundaries of the human body, or should it consist solely in some minute inner action of choosing? (E.g. Goldman (1971) and perhaps Wittgenstein (1969/2003, sect. 402 and 476) at the former extreme, and Hornsby (1980) at the latter).
45. Cf. Dancy (2003) and Davis (2003).
46. Hempel (1948, 1965) argued that causal explanations must be lawlike or assume laws and this view has been very influential. It is of note that Hempel was plagued by the problem of how to apply his nomological account of explanation to explanation in terms of reasons.
47. Cf. Davidson (1970/2001b).
48. Cf. Davidson (1963/2001a) and (1970/2001b) respectively.
49. Cf. Schuler (2003, ch. 1).
50. Cf. Cullity and Gaut (1997).
51. This does not stand in conflict with the position, defended in the preceding chapter, that there can only be one rationality and agents must be by and large rational. Because, as just noted, that requirement is only a 'by and large' requirement whereas lawlikeness is an all or nothing type of requirement.
52. Cf. Cain (2003) and Alvarez (2009b).
53. This is to deny that we can have fully determinate counterfactual descriptions of the relations between mental states and actions. This goes against Hornsby's suggestion that the holding of counterfactuals shows that causation is at issue (e.g. Hornsby (1993)). Given the popularity of counterfactual accounts of causality (e.g. Lewis (1973)), this also speaks against remarks such as Sosa's (1993) that it is not clear that causation must go hand in hand with lawlikeness.
54. Cf. Gopnik (1997) and Goldman (1989) respectively.

References

Alvarez, M. (2009a). How Many Kinds of Reasons? *Philosophical Explorations, 12*, 181–193.

Alvarez, M. (2009b). Actions, Thought-Experiments, and the 'Principle of Alternate Possibilities'. *Australasian Journal of Philosophy, 87*, 61–81.

Alvarez, M. (2010a). *Kinds of Reasons: An Essay in the Philosophy of Action*. Oxford: Oxford University Press.

Alvarez, M. (2010b). Reasons for Action and Practical Reasoning. *Ratio, 23*, 355–373.

Anscombe, G. E. M. (1957/2000). *Intention*. London: Harvard University Press.

Aristotle. (384-22BC/2001). Nicomachean Ethics (Ethica Nicomachea) (W. D. Ross, Trans.). In R. McKeon (Ed.), *The Basic Works of Aristotle* (pp. 935–1112). New York: The Modern Library (Random House).

Aristotle. (384-22BC/2006). The Movement of Animals (De Motu Animalium) (E. S. Foster, Trans.). In *Aristotle XII: Parts of Animals, Movement of Animals and Progression of Animals* (pp. 433–479). London: Harvard University Press (Loeb Library).

Audi, R. (1989). *Practical Reasoning*. London: Routledge.

Beanblossom, R. E. (1971). Walton on Rational Action. *Mind, 80*, 278–281.

Ben-Yami, H. (1997). Against Characterizing Mental States as Propositional Attitudes. *The Philosophical Quarterly, 47*, 84–89.

Bird, A. (1998). Dispositions and Antidotes. *The Philosophical Quarterly, 48*, 227–234.

Broome, J. (2001). Normative Practical Reasoning. *Proceedings of the Aristotelian Society, Supplementary Volume, 75*, 175–193.

Broome, J. (2002). Practical Reasoning. In J. Bermudez & A. Millar (Eds.), *Reason and Nature: Essays in the Theory of Rationality* (pp. 85–111). Oxford: Oxford University Press.

Broome, J. (2009). Motivation. *Theoria, 75*, 79–99.

Brown, R., & Ladyman, J. (2009). Physicalism, Supervenience, and the Fundamental Level. *The Philosophical Quarterly, 59*, 20–38.

Cain, J. (2003). Frankfurt Style Examples. *Southwest Philosophy Review, 19*, 221–229.

Cameron, R. (2008). Turtles All the Way Down: Regress, Priority, and Fundamentality. *The Philosophical Quarterly, 58*, 1–14.

Cullity, G., & Gaut, B. (1997). Introduction. In G. Cullity & B. Gaut (Eds.), *Ethics and Practical Reason* (pp. 1–27). Oxford: Oxford University Press.

Dancy, J. (2000). *Practical Reality*. Oxford: Oxford University Press.

Dancy, J. (2003). Replies. *Philosophy and Phenomenological Research, 67*, 468–490.

Davidson, D. (1963/2001a). Actions, Reasons, and Causes. In *Essays on Actions and Events* (pp. 3–19). Oxford: Oxford University Press.

Davidson, D. (1970/2001b). Mental Events. In *Essays on Actions and Events* (pp. 207–225). Oxford: Oxford University Press.

Davidson, D. (1973/2001c). Freedom to Act. In *Essays on Actions and Events* (pp. 63–81). Oxford: Oxford University Press.
Davis, W. A. (2003). Psychologism and Humeanism. *Philosophy and Phenomenological Research, 67*, 452–459.
Frankfurt, H. G. (1969). Alternate Possibilities and Moral Responsibility. *The Journal of Philosophy, 66*, 829–839.
Goldman, A. I. (1971). The Individuation of Action. *The Journal of Philosophy, 68*, 761–774.
Goldman, A. I. (1989). Interpretation Psychologized. *Mind and Language, 4*, 161–185.
Gopnik, A. (1997). The Scientist as Child. *Philosophy of Science, 63*, 485–514.
Hempel, C. G. (1948). Studies in the Logic of Explanation. *Philosophy of Science, 15*, 135–175.
Hempel, C. G. (1965). Aspects of Scientific Explanation. In *Aspects of Scientific Explanation and Other Essays in the Philosophy of Science* (pp. 331–496). New York: The Free Press.
Hornsby, J. (1980). *Actions*. London: Routledge and Kegan Paul.
Hornsby, J. (1993). Agency and Causal Explanation. In J. Heil & A. Mele (Eds.), *Mental Causation* (pp. 161–188). Oxford: Oxford University Press.
Hume, D. (1739–40/1985). *A Treatise of Human Nature*. London: Penguin.
Kant, I. (1787/1996). *Critique of Pure Reason* (W. S. Pluhar, Trans.). Indianapolis: Hackett.
Korsgaard, C. M. (2008). Acting for a Reason. In *The Constitution of Agency: Essays on Practical Reason and Moral Psychology* (pp. 207–209). Oxford: Oxford University Press.
Korsgaard, C. M. (2009). The Activity of Reason. *Proceedings and Addresses of the American Philosophical Association, 83*, 23–43.
Leibniz, G. W. F. von. (1714/1974). The Monadology (G. Montgomery, Trans., revised by A. R. Chandler). In A. R. Chandler (Ed.), *The Rationalists* (pp. 455–471). London: Anchor Books.
Lewis, D. K. (1973). Causation. *Journal of Philosophy, 70*, 556–567.
Lowe, E. J. (2008). *Personal Agency*. Oxford: Oxford University Press.
Manson, N. C. (2004). Reason Explanation a First-Order Rationalizing Account. *Philosophical Explorations, 7*, 113–129.
Martin, C. B. (1994). Dispositions and Conditionals. *The Philosophical Quarterly, 44*, 1–8.
McGinn, C. (1979). Action and Its Explanation. In N. Bolton (Ed.), *Philosophical Problems in Psychology* (pp. 20–42). London: Methuen.

Mele, A. R. (2003). *Motivation and Agency*. Oxford: Oxford University Press.
Nagel, T. (1978). *The Possibility of Altruism*. Chichester: Princeton University Press.
Parfit, D. (1987). *Reasons and Persons*. Oxford: Oxford University Press.
Parfit, D. (2001). Rationality and Reasons. In D. Egonsson, B. Petterson, & T. Ronnow-Rasmussen (Eds.), *Exploring Practical Philosophy: From Action to Values* (pp. 17–39). Aldershot: Ashgate.
Platt, M. d. B. (1979). *Ways of Meaning*. London: Routledge.
Rundle, B. (1997). *Mind in Action*. Oxford: Oxford University Press.
Schueler, G. F. (2003). *Reasons and Purposes*. Oxford: Oxford University Press.
Smith, M. (1987). The Humean Theory of Motivation. *Mind, 96*, 36–61.
Smith, M. (1994). *The Moral Problem*. Oxford: Blackwell.
Smith, M. (2003). Humeanism, Psychologism, and the Normative Story. *Philosophy and Phenomenological Research, 67*, 460–467.
Sosa, E. (1993). Davidson's Thinking Causes. In J. Heil & A. Mele (Eds.), *Mental Causation* (pp. 41–50). Oxford: Oxford University Press.
Thagard, P. R. (2006). Desires are Not Propositional Attitudes. *Dialogue, 45*, 151–156.
Wallace, R. J. (2003). Explanation, Deliberation, and Reasons. *Philosophy and Phenomenological Research, 67*, 429–435.
Walton, K. A. (1967). Rational Action. *Mind, 76*, 537–547.
Williams, B. (1995). Acts and Omissions, Doing and Not Doing. In *Making Sense of Humanity: And Other Philosophical Papers 1982–1993* (pp. 56–64). Cambridge: Cambridge University Press.
Wittgenstein, L. (1969/2003). *On Certainty* (D. Paul & G. E. M. Anscombe, Trans.). Oxford: Blackwell.

6

Rationality

In the last chapter I showed that acting for a reason involved being aware of a reason, and then choosing to act for that reason, adopting that action as a goal, and acting. For example, I gain pleasure from whistling and I might become aware of this fact and hence form a desire to whistle, decide to whistle, and whistle. This shows that there is more to acting for a reason than simply acting in a way that accords with a reason. The present chapter is concerned with rationality, or rational actions. It appears that this takes us beyond the topic of simply acting for a reason, because an agent can act rationally when there is no reason for that action. If I am not being approached by a speeding car, but I believe that I am, it appears to be rational for me to jump out of the road. These error cases have led to the suggestion that a rational agent must meet a number of complex normative requirements.

Thus it appears that we can distinguish two kinds of normative element. The first stems from reasons. There are actions that one has a reason to perform—reasonable actions. The second concerns requirements of rationality. It concerns actions that it is rational to perform—rational actions. It appears that rational actions will not always be reasonable actions. This has led some to believe that the requirements of rationality

Olley (F.O.C.H.) Pearson, *Rationality, Time, and Self*,
https://doi.org/10.1007/978-3-319-71973-3_6

provide a source of normative requirements independent of reasons. In this chapter I will argue that this view is mistaken.

There are only reasons to act and no independent requirements of rationality. Rationality is primarily a matter of responding correctly to reasons. It is a matter of acting for j-reasons. If a rational action is not a reasonable one, this can only be because the agent is mistaken about what j-reasons obtain. The rationality in this error case is derived from the rationality of the primary case. The error case qualifies as rational because of similarities it bears to the primary case of acting for a j-reason. Sensitivity to reasons in the sense of knowing that they obtain is not a rational matter. A rational agent cannot be expected to be omnipotent. However, sensitivity to reasons in the sense of reacting appropriately to those reasons one is aware of is a matter of rationality. When an agent acts rationally and acts for a j-reason that agent's j-reason and action form a structure. This structure is isomorphic with the structure formed by the agent's belief about the j-reason and the agent's desire which gives his or her goal in acting. When an agent is mistaken about what reasons obtain and therefore fails to act for a j-reason the agent may nonetheless still have a structure of belief and goal giving desires. If this structure would have been isomorphic with a structure formed by the agent's action and a j-reason had the belief been true, then the agent's action is sufficiently similar to the primary case of acting for a reason to qualify as a rational action. If an agent were sufficiently knowledgeable, then his or her rational actions would simply be his or her reasonable actions and the agent would be rational simply because the agent was reasonable. If an agent acts rationally but not for a j-reason, then the agent is not doing something that the agent ought to do though the agent is sufficiently close to this ideal to be in some sense excused.

6.1 Distinction Between Reasonable & Rational Actions

If I believe that eating salmon gives me pleasure, then it is rational for me to eat salmon.[1] This is so whether or not eating salmon does actually give me pleasure and thereby whether or not I have a reason to eat salmon. This shows that rational and reasonable actions do not need to coincide.

Rationality thus appears to track the reasons one believes oneself to have for an action rather than the reasons one does actually have.[2] If eating salmon would give me pleasure I would have a reason to eat salmon and it is because of this that my eating salmon is a rational action.

As already noted, it is popular to suppose that the conclusion of a piece of practical reasoning is a desire or intention. It is also popular to suppose, and I take this to be correct, that the conclusion of a piece of theoretical reasoning is a belief. (The distinction drawn between practical and theoretical reasoning is often taken to rests on this distinction of conclusions.[3]) With this in mind rationality has often been described as a matter of the consistency between the states of an agent. Specifically, it concerns the consistency between an agent's beliefs about what reasons obtain and the agent's other beliefs or intentions formed for those reasons or chosen in response to those beliefs. Returning to my example, it might be said that rationality concerns the consistency between my believing that I will gain pleasure from eating salmon and my intending to eat salmon.

I believe that the conclusion of a piece of practical reasoning is a chosen action not an intention. Nevertheless, when an agent chooses to act the agent will have that action as a goal: the agent will desire to perform that action. This desire can take the role of the intentions just alluded to so that rationality can be thought of as a matter of the consistency between an agent's beliefs about what reasons obtain and that agent's other beliefs and the desires which give the agent's goals in acting.

The consistency at issue is often taken to be given by a number of complex normative requirements (CNRs). For example, in the case of my eating salmon the requirement at issue appears to be of the form:

> CNR1: An agent ought rationally to have *F* as a goal if that agent believes there is all things considered reason for him/her to *F*.[4]

The notion of all things considered reason is used to make matters more precise because for most actions an agent will be aware of numerous reasons for and against that action. In a circumstance in which an agent is aware of a reason to *F*, but also of more reasons not to *F*, then *F*ing will not be the rational thing to do. So, in giving my salmon example I supposed there were no other relevant reasons at issue.

When I say 'all things considered reason' this does not mean that the agent has considered everything. Rather, I mean that if all of the reasons that the agent took to obtain did obtain, then the agent would have most reason to *F*. The notion of most reason is not simply a matter of having a greater number of reasons to *F*. It is a matter of having a greater weight of reasons to *F*. I am thus speaking of *pro tanto* reasons which can conflict with one another and be weighed against one another.[5] I will not go into the issue of how reasons ought to be weighed but I take the notion to have a familiar sense. For example, people often take the fact that an action will be very good for their health to be a weightier reason to act than the fact that an action will give them a very small amount of pleasure. The weight that an agent gives to reasons can be revealed in their choosing.[6] We can ignore the complexities of weighing, as I did in giving my example, if we suppose the agent only believes there is one reason to *F* and doesn't believe there are any reasons not to *F*. In such a case the agent will believe there is all things considered reason to *F*. Correspondingly, we ought to say that a reasonable action is one that there is most reason to perform not simply a reason to perform.[7]

The rationality of an agent—or the relevant consistency of their states—appears to involve meeting a number of other requirements too:

> CNR2: An agent ought rationally to have *G* as a goal if that agent has *F* as a goal and that agent believes that he/she cannot *F* unless he/she *G*s.[8]
>
> CNR3: An agent ought rationally not to have *F* as a goal if that agent believes he/she cannot *F* and *G*, and has *G* as a goal.[9]
>
> CNR4: An agent ought rationally to believe that *p* if that agent believes that *q* and believes that *q* entails *p*.[10]

Perhaps the wording or detail of these CNRs will be disputed, but they clearly point towards something plausible and are precise enough for my purposes. It might be possible to add more CNRs to this list,[11] but my intention here is merely to emphasize the extent to which rationality appears to be given by CNRs and how this differs from reasonableness.

Variants of CNR2 are often referred to as the *Instrumental Principle* because they concern taking means to ends. Suppose that I believe that I have all things considered reason to go walking and that I believe that putting on my shoes is a necessary means to going walking, and this is something I am quite capable of doing. I would be irrational if I did not put on my shoes. The point of CNR2 is to capture this irrationality of being aware of means to our ends but being unmoved to take those means. This irrationality lingers even if we suppose that I was in fact mistaken and I have no reason to go walking. So reasonableness and rationality appear to part company.

If I decide to go to the cinema and know that I cannot go to the cinema and to the restaurant, then I clearly should not also decide to go to the restaurant. (I should not decide this and keep my former decision.) Someone who acted in this way would be irrational and this is the point of CNR3. Again this irrationality appears to linger however we suppose the reasons to actually fall. It lingers whether or not I have equal reason to do both or more reason to do one than the other.

CNR4 probably needs to be tightened up so that it is clear that the agent believes *q* and that *q* entails *p* at the same time. We might also want to say that the entailment is in some way apparent and not merely borne out by a long series of connectable premises. But the point of the CNR is clear. If I believe that I am in Durham and believe that if I am in Durham I am not in London, then it is quite irrational of me to remain uncertain as to whether or not I am in London. Moreover, this remains the case even if we suppose that I am mistaken and that I am in fact in London. I might then have no reason to believe that I am not in London, but it would be irrational for me not to do so.

Rationality and reasons thus appear to be quite distinct because an action can be rational but not reasonable. Moreover, the two appear to be distinct in form too because whether or not an action or belief is reasonable depends only on the reasons—or facts—that obtain. However, whether or not a belief or goal giving desire is rational does not depend on the facts as much as on what the agent believes to be the case. Further, rationality appears to be determined by requirements of consistency—CNRs.

6.2 Rational Obligations

In stating the requirements of rationality—or CNRs—I spoke of what an agent ought rationally to do as if rationality can place obligations on an agent. The examples I mentioned support this idea: apparently an agent should avoid being irrational. It is thus plausible to think that CNRs provide agents with j-reasons.[12]

However, the idea that the requirements of rationality provide agents with j-reasons quickly runs into a familiar problem. Suppose that CNR1 provides me with a j-reason to act in accord with it. We saw that if I believe that eating salmon will give me pleasure, then according to CNR1 I ought to adopt eating salmon as a goal. But, if uninterrupted, my adopting eating salmon as a goal will simply be a matter of my eating salmon. So, in short, I ought to eat salmon. This is so whether or not eating salmon will give me pleasure, or whether or not I have any other j-reasons to eat salmon (excluding CNR1 itself). It follows that if I have a j-reason to accord with CNR1, then simply through forming the belief that a j-reason obtains I bring it about that I have a j-reason to eat salmon. But this is exactly the sort of scenario that we considered in Chap. 4 and that we there rejected. One cannot simply bootstrap j-reasons into existence by forming beliefs or desires. It does not follow from the fact that I happen to form the belief that I have a j-reason to *F* that I do in fact have a j-reason to *F*. It remains the case that I will not gain pleasure from eating salmon and that I have no j-reason to do so or to adopt doing so as a goal even if I mistakenly believe that I do. Beliefs about j-reasons are not self-fulfilling.

As we saw in Chap. 4, this problem has been considered by a number of writers and a solution has been proposed.[13] As it stands one could read CNR1 in two ways:

> CNR1W: An agent ought rationally to <have *F* as a goal if that agent believes there is all things considered reason for him/her to *F*>.
>
> CNR1N: An agent ought rationally to <have *F* as a goal> if that agent believes there is all things considered reason for him/her to *F*. [Or: If an agent believes there is all things considered reason for him/her to *F*, then he/she ought rationally to <have *F* as a goal>.]

In CNR1W the conditional falls within the scope of the rational obligation, whereas in CNR1N it does not. (The former is a *wide scope* requirement and the latter is a *narrow scope* requirement.) This makes a difference to how one can satisfy the CNR.

Suppose that I believe that eating salmon will give me pleasure. In order to satisfy CNR1N I would have to eat salmon. However, I could satisfy CNR1W without eating salmon and by instead dropping my belief that eating salmon will give me pleasure. It follows that if we interpret CNR1 with wide scope—CNR1W—then we do not have the bootstrapping problem. Rather, if one forms the belief that one has a reason to *F*, CNR1W gives one a reason to either *F* or drop one's belief, but it does not give one a reason to adopt either one of these disjuncts in particular.

Similar remarks hold for the other CNRs. Each is susceptible to a bootstrapping criticism, but this criticism can be rebutted if it is insisted the CNRs are to be understood in a wide scope manner which denies we can derive a specific obligation/reason from a reason to satisfy a complex. So it appears that the requirements of rationality do give agents j-reasons, and that we can make sense of this fact.

6.3 The Denial of Rational Obligations

As just noted, accepting that CNRs provide agents with j-reasons is dependent upon denying detachment because otherwise we end up bootstrapping j-reasons into existence. However, in this section I will argue that we cannot deny detachment in the case of CNRs and therefore that we should not take them to provide reasons (my argument here will largely follow Kolodny (2005, 2007a, 2007b, 2008)).

Kolodny points out that we can interpret CNRs as either state or process requirements. A state requirement simply demarcates certain states or combinations of states as irrational. A process requirement tells an agent that the agent ought to do certain things or that certain processes are irrational. Requirements of rationality appear to be process rather than state requirements because they can be used to advise or guide agents and not simply to evaluate states agents are in. If they are reasons, or provide reasons, then they clearly provide reasons to do certain things not simply to be a certain way.

Broome (2007b) and Kolodny (2007a) have both noted that an agent will only violate a wide scope state requirement if they would also have violated a narrow scope one, and vice versa. Consider CNR1 as a state requirement:

> CNR1WS: An agent ought rationally to <have *F* as a goal at time *t* if that agent believes at time *t* that there is all things considered reason for him/her to *F*>.
>
> CNR1NS: An agent ought rationally to <have *F* as a goal at time *t*> if that agent believes at time *t* that there is all things considered reason for him/her to *F*.

There are three ways that an agent might be relevant to this requirement: A1, the agent has *F* as a goal at *t* and believes at *t* that there is all things considered reason for him or her to *F*; A2, the agent believes at *t* that there is all things considered reason for him or her to *F* but the agent does not have *F* as a goal at *t*; A3, the agent does not believe that he or she has all things considered reason to *F*.

If the agent is in state A1, then the agent will satisfy both CNR1WS and CNR1NS. If the agent is in state A2, then the agent will violate both CNR1WS and CNR1NS. And, if the agent is in state A3, then the agent will satisfy CNR1WS and will not have any relevance to CNR1NS. This makes it appear that as far as state requirements are concerned there is in fact little difference between CNR1WS and CNR1NS. Both evaluate the same states to be irrational.[14] (And we shouldn't focus instead on the positive ascription of rationality based on fulfilling such requirements, rather than the negative one of not breaking them, precisely as it is not less rational to fail to fulfil an irrelevant CNR.)[15]

However, if we take CNR1 to be a process requirement then the distinction between a wide and narrow scope interpretation is much more telling:

> CNR1WP: An agent ought rationally if that agent believes at time *t* that there is all things considered reason for him/her to *F* and he/she does not

at *t* have *F* as a goal, to <either going forward from *t* on the basis of the content[16] of that belief take up *F* as a goal, or going forward form *t* on the basis of the content of the lack of his/her having *F* as a goal revise his/her belief that there is all things considered reason to *F* >.

CNR1NP: An agent ought rationally if that agent believes at time *t* that there is all things considered reason for him/her to *F* and he/she does not at *t* have *F* as a goal to <going forward from *t* on the basis of the content of that belief take up *F* as a goal >.

These CNRs differ from the original CNR1 but the differences are there to make best sense of the requirement being understood as a process requirement. As a state requirement CNR1 picks out a certain combination of states as irrational. The process requirement reflects this by considering processes that will avoid such a state. If we drop the 'on the basis of' condition, then the requirement would clearly be false because it is not rational to make that move in all possible ways. It is not rational if one drops a belief or gains a desire as a result of being hit on the head. A different way to fill out this condition would be to say that an agent ought to do so on the basis of other relevant beliefs or goals, or lacks thereof. But, this is no good because there needn't be any other beliefs or goals or lacks thereof of this sort.[17]

An agent might violate CNR1NP but not violate CNR1WP. This will happen if an agent believes at *t* that there is all things considered reason for him or her to *F* and he or she does not at *t* have *F* as a goal, and going forward from *t* he or she does not take up *F* as a goal, but he or she does drop his or her belief that there is all things considered reason for him or her to *F*. This indicates that a process understanding of CNR1 is different from a state one, and moreover, that if one thinks that wide scope and narrow scope requirements say different things of rationality, as it is common and natural to, then one must be thinking of process requirements.

Now it is clear that requirements of rationality are process requirements it is no longer clear that we ought to take them as wide scope requirements. It is rational to move as CNR1NP prescribes from believing oneself to have all things considered reason to *F* to having *F* as a goal. CNR1WP, although not prescribing this behaviour, nonetheless recommends it as rational. However, CNR1WP also recommends as rational the quite different behaviour of moving from lacking *F* as a goal[18] to the abandonment of a belief that one has all things considered reason to *F*. But this latter behaviour does not appear to be rational. Rather, such behaviour strikes one as quite wrong. What one ought to do shouldn't be expected to bend to one's desires.

Similar remarks apply to CNR4. It is irrational to move from a lack of belief that *p*[19] to the abandonment of a belief that *q* or that *q* entails *p*. In fact, it is hard to see how the lack of a belief or goal—a lack as opposed to the fact that a belief or goal is lacked—could be effective at all. (This is not to say that grounds for believing not-*p* cannot be grounds for believing not-*q* or not-*q*-entails-*p*. These grounds are not what are discussed by these CNRs.) Further, the problems here would apply equally to various different interpretations of the CNRs I give. The point of CNRs like 1 and 4 is to link beliefs about reasons (or possible reasons) to further—concluding—beliefs or goal giving desires which might be justified by those reasons (were they to obtain). The CNRs in this way reflect the structure of reasons (and the would-be structure of would-be reasons). If the reasons believed to obtain did obtain, the concluding beliefs and goal giving desires would be justified. The denial of detachment aims precisely to allow that the beliefs about what reasons obtain might be mistaken and hence that it might be that they ought to be changed. The linking of this with the concluding beliefs or goal giving desires in a CNR entails that this change is directly linked to the possession or lack thereof of these concluding states. These points must be common to the variety of versions of the CNRs 1 and 4. The former is necessary to avoid bootstrapping. The latter is necessary as the CNRs merely concern relations between states. But, these points are all that are required to apply the above criticism.

It will not do to respond to this by suggesting that the obligations of rationality are distinct from j-reasons so that we have two forms of broadly rational oughts. It is obscure what these other rational obligations might be and how they would interact with the obligations of j-reasons. Further, this would not seem to avoid the troubles raised here because the bootstrapping argument could be applied in a similar way to rational obligations that were not reasons. It is not simply the case that forming a belief that one has reason to *F* or a desire to *F*, does not give one a j-reason to *F*. It is the case that such beliefs and desires do not make it the case that one ought to *F* whether or not we interpret this ought as grounded in j-reasons.

If we suppose that the requirements of rationality provide reasons or obligations, we ought to take many of these obligations to be narrow scope to avoid endorsing irrational behaviour. But, if we take the obligations to be narrow scope, then we can simply bootstrap j-reasons (or oughts) into existence by believing that there are j-reasons. But, this bootstrapping is unacceptable. Therefore, we should not suppose that requirements of rationality provide reasons or obligations

6.4 Explaining the Appearance of Rational Obligations

Kolodny (2005, 2007b, 2008), following Raz (2005), has offered an error theory for why one might have thought CNRs did give rational obligations. His method has two central stages. Firstly, it is noted that people might be motivated to think that CNRs provide obligations because they think that if they are in a state/process that would violate a CNR, then they are in a state/process that would violate an obligation. So for each CNR a relevant violation claim is drawn out. Secondly, it is shown that reasons could provide the obligations that would be violated. It is thus shown that a state/process that would be against a CNR would also be against reasons. This shows that we do not need to refer to CNRs to provide obligations. Moreover, this closeness between the nature of reasons and the CNRs also shows why people might mistakenly take CNRs to provide obligations.

In the case of CNR2, for example, the violation claim appears to be:

> VC2: If an agent has *F* as a goal and believes that *G* is a necessary means to *F*, and that agent does not have *G* as a goal, then that agent violates some obligation.

If CNR2 provided an obligation, then this would be violated in the case picked out by the violation claim. However, we needn't adopt the CNR on these grounds because we could refer to the breaking of different obligations to satisfy the violation claim. Specifically, in line with the argument above that only reasons provide obligations, we can refer to a relevant reasons claim:

> RC2: If an agent has conclusive reason to believe that he/she will *F* only if he/she *G*s, then that agent has reasons to *G* at least as strong as his/her reasons to *F*.[20]

According to RC2 an agent will go against reasons if that agent has *F* as a goal, believes that *G* is a necessary means to *F*, and does not have *G* as a goal. That is, the nature of reasons picked out by RC2 shows that reasons can provide the obligations that VC2 refers to (CNR2 is not necessary). Moreover, RC2 picks out what is plausible in VC2. Either an agent needn't *G* to *F* or if that agent has reason to *F* that agent has reason to *G*.

One should not be worried about the talk of conclusive reasons in RC2. We can say two things of the case in which an agent has a merely inconclusive reason for his or her beliefs. Firstly, we can say the case for conclusive reasons itself provides an error theory for the case for inconclusive reasons: because of the violation of reasons in one case it is naturally assumed there is a violation of reasons in the other case. Secondly, we can note that as an agent's belief becomes less conclusive, and so less certain, it is less clear that the agent is irrational in failing to act in accord with it. If I am uncertain I have a reason to do something, then I am not obviously irrational for not doing it.

One might have been mistakenly led to believe that CNR2 provides one with obligations because one accepted VC2. However, VC2 in fact has an explanation stemming from reasons which is revealed by RC2. Thus reasons provide the only rational obligations at issue, but it is understandable how one might have mistakenly thought that CNRs were of relevance as both would explain VC2.

This pattern can be repeated for the other CNRs I mentioned above. In the case of CNR1 we might refer to:

VC1: If an agent believes that there is all things considered reason to *F* and does not have *F* as a goal, then that agent violates some obligation; and,

RC1: If an agent has conclusive reason to believe that there is all things considered reason to *F*, then that agent has reason to adopt *F* as a goal.

In the case of CNR3, we might refer to:

VC3: If an agent believes that he/she cannot *F* and *G*, and he/she has *F* as a goal and has *G* as a goal, then that agent violates an obligation; and,

RC3: If an agent believes that he/she cannot *G* and *F*, has *G* as a goal, and has *F* as a goal, then either that agent believes without conclusive reason that he/she will *G*, believes without conclusive reason that he/she will *F*, or believes without conclusive reason that he/she cannot *G* and *F*.[21]

The connections in RC3 are made in part by assuming that if an agent adopts an action as a goal and so chooses to do something, then that agent will believe that he or she will do it. An agent cannot choose to do something that that agent believes he or she will not do.[22] So if the agent has both *F* and *G* as goals, then that agent must believe he or she will *F* and *G*, and one of these beliefs must be wrong if the agent correctly believes he or she cannot *F* and *G*. Finally, in the case of CNR4 we might refer to:

VC4: If an agent believes that *q* and that *q* entails *p*, and that agent does not believe that *p*, then that agent violates an obligation; and,

RC4: If an agent lacks conclusive reason to believe that *p*, then that agent either lacks conclusive reason to believe that *q* or lacks conclusive reason to believe that *q* entails *p*.

Of course much more could be said about these RCs and VCs.[23] However, my point here has merely been to show that it is possible to provide a plausible error theory to account for why it is that someone might take CNRs to provide rational obligations when in fact they do not. The individuals at issue no doubt accept a number of VCs related to these CNRs and their mistake arises from supposing that the CNRs account for the VCs when in fact it is RCs that fulfil this role. This moves us towards the idea that rationality is little more than responsiveness to reasons. One is irrational when not appropriately responsive to reasons.

6.5 What Rationality Consists In

The preceding discussion shows that if we focus on the cases in which an agent is mistaken about what reasons obtain we can easily be misled and mistakenly conclude that there must be requirements of rationality distinct from the requirements of reasons. There is a similarity here with the reasons case discussed in Chap. 4. When discussing reasons we also saw that we should not let appearances concerning error cases mislead us but should instead focus on the good cases and see how the two relate.

We will get a better account of rationality if we focus on the primary or paradigm cases of rational actions and rationally forming beliefs or goal giving desires. I will focus on the case of rational actions and practical rationality. In an ideal case an agent will be aware of all the reasons there are and that agent will do the action that there is most reason to do for the reasons that there are to do it. The rational action will be the action that the agent ought to perform and this obligation will stem from the reasons that there are to perform it. However, agents are not omniscient and they should not be expected to be to be rational. So, we should consider a paradigm case of acting rationally rather than an ideal case. A paradigm case of acting rationally is a case in which an agent has an all things considered j-reason to do something and that agent does that thing for that reason. The agent is ignorant of some reasons but acts to do what the reasons he or she is aware of oblige and for those reasons. If I will gain pleasure from eating salmon, I am aware of that, and I eat salmon for that reason, then I will be acting rationally. In the ideal and paradigm cases rationality is simply a matter of doing what the j-reasons one is aware of prescribe doing and doing it for those reasons. In the paradigm cases acting rationally just is acting for reasons once we allow that omnipotence isn't a requirement of rationality. With the paradigm case in mind there is little temptation to say that one is obliged to act rationally. It is obvious that acting rationally just is doing what one is obliged by the reasons to do.

We should not think of there being additional requirements of rationality or requirements beyond j-reasons when we turn to consider error cases in which an agent mistakenly takes j-reasons to obtain that do not obtain. Rather, we should say that an agent can sometimes be forgiven for not acting for a j-reason. In an error case an agent does not do what that agent ought to do because that agent does not act for a j-reason. When we nonetheless describe that agent as rational this is not because that agent has done what he or she ought to do. (It is not because the agent has fulfilled a rational requirement like a CNR.) We describe that agent as rational because what that agent does is sufficiently close to the paradigm case for the agent to be excused for falling short and failing to do as he or she ought to. If I mistakenly believe that I will gain pleasure from eating salmon and have a goal giving desire to eat salmon, then I am rational. This is not because I act for a reason nor because I fulfil a CNR. It is because I am sufficiently close to the case of acting for a j-reason to be forgiven.

More specifically, we can say that an agent will be forgiven if, had that agent's beliefs about what reasons obtained been true, these beliefs and other beliefs and goal giving desires of the agent would form a network isomorphic with that of the network of reasons and the beliefs and actions they are reasons for. For example, if I will gain pleasure from eating salmon, there is no reason for me not to eat salmon, I am aware of this reason to eat salmon, and I eat salmon for this reason, then I will be acting rationally. My action will be a paradigm of rationality because it will be a case of acting for an all things considered j-reason. In such a case my reason for eating salmon and my eating salmon are related in a particular way. Further, my belief about that reason and my goal giving desire of eating salmon are also related in an isomorphic way. Suppose instead that I would not gain pleasure from eating salmon so I had no reason to do so, but I believed I would and had eating salmon as a goal and resultantly ate salmon. In such a case my belief and goal giving desire would not form a structure isomorphic with the structure formed by my j-reasons and my actions because there were no j-reasons. However, if my beliefs had been true, there would have been j-reasons and there would have been isomorphism. I suggest that something like this demarcates those cases which qualify as being rational from those which do not and it does so because it demarcates a number of cases in which we would forgive an agent for not acting for an all things considered j-reason. These are perhaps cases in which one is tempted to say that if the agent's beliefs about what reasons obtained had been true, that agent would have acted [formed a belief or goal giving desire] for all things considered j-reasons. The obtaining of this counterfactual conditional would encourage forgiveness and thereby qualify the agent as rational.

Similar points are made by both Raz and Kolodny.[24] Raz (2005) speaks of a faculty of rationality which enables an agent to respond appropriately to reasons. For Raz, when an agent acts irrationally this is not a matter of that agent going against particular reasons or of that agent breaking a CNR. It is a matter of the agent acting in a way that reveals a fault in the agent's rational faculty. The fault is an insensitivity to the general nature of reasons, where a sensitivity to reasons in the relevant sense would be captured by the isomorphism I spoke of just above. However, I find this notion of a faculty of rationality unclear and even misleading insofar as it suggests that one has a sensitivity to reasons in the error cases where there are in fact no reasons to be sensitive to. I take the more telling issue to be the closeness of the relation between error cases and paradigm cases.

Kolodny again draws the boundary between rational and irrational cases in the same place, though he focuses on the perspective of the agent and the reasons that the agent takes himself or herself to have. For Kolodny, an agent will be rational insofar as that agent acts in accord with the reasons that the agent takes himself or herself to have to act. When I mistakenly believe that I will gain pleasure from eating salmon from my perspective I have a reason to eat salmon and so I will be rational so long as I do so. An advantage of this view is that it can explain why, from the agent's perspective, being told what the rational thing to do is not merely a matter of having the situation evaluated, but it is also a matter of being given guidance about what to do. From the third party's perspective describing such an action as rational is merely evaluating it, but from the perspective of the agent mistaken about what reasons obtain this is a matter of being told what these reasons are reasons for and hence to do. Thus from the perspective of the agent the appearance of obligations lingers in the error cases even though there are in fact none. I have some sympathy with Kolodny about this, though it remains true that these error cases qualify as rational solely on the basis of the closeness they bear to paradigm cases. If acting rationally did not involve acting for reasons as it does in the paradigm cases, it would not involve acting in accord with the reasons one takes oneself to have in the error cases.

This account clearly links to the patterns suggested by CNRs 1 and 4. In fact, satisfying CNRs 1 and 4 involves having a network of beliefs about reasons[25] and other beliefs and goal giving desires which, if those beliefs about reasons were true, would be isomorphic with the network of j-reasons and the beliefs or actions they are j-reasons for. The story with the other CNRs is perhaps less clear, but the RCs noted above open a window here. The RCs show that in paradigm cases of acting (desiring or believing) rationally in accord with these CNRs one is acting (desiring or believing) in accord with reasons. Acting (desiring or believing) rationally is also arguably a matter of acting (desiring or believing) in a way somehow responsive to reasons or one's beliefs about these reasons. Only those bodily movements with a specific form of responsiveness are rational actions. It follows that in the paradigm cases of acting (desiring or believing) in accord with these CNRs an agent will not merely act (desire or believe) in accord with reasons but will act (desire or believe) for reasons. The error cases will then have a closeness to these paradigm cases which can be described in terms of an isomorphism or would be isomorphism, even if this now concerns the states mentioned by the CNR and other states, rather than merely the states mentioned by the CNR as with CNRs 1 and 4.

Irrationality thus consists in acting without appropriate responsiveness to reasons. There are two obvious ways that this can come about. If an agent is aware of a number of reasons to perform an action, but a number of far stronger reasons not to perform it, and yet nonetheless performs the action, then that agent does not have the appropriate responsiveness to reasons and is irrational. Alternatively, if an agent is aware of a fact and for that fact performs an action, but that fact is not a j-reason for the action, then the agent does not have appropriate accord with reasons and is therefore irrational.

6.6 Summary

The only rational obligations on an agent are those provided by j-reasons. In paradigm cases an agent will be aware of what there is most reason to do and the agent will do that thing for those reasons. However, an agent can be mistaken about what reasons obtain. When an agent is mistaken about this that agent can still qualify as rational. The agent will not qualify as rational because the agent does as the agent ought to or because the agent fulfils a CNR. The agent will qualify as rational because the agent acts in a way sufficiently close to the paradigm case. Specifically, because the agent has a network of beliefs and goal giving desires which would be isomorphic with the network of reasons and the actions they are reasons for if the agent's beliefs about what reasons obtained were true. From the agent's perspective the rational thing to do will always appear to be the thing that the agent ought to do. Though a third party can see that in error cases what it is rational for an agent to do and what they ought to do, or the actions that they have reasons to perform, can be distinct. But, error cases and paradigm cases will always be similar in a way that shows the rationality of the former derives from the reasonableness of the latter. This links back to the account of acting for a reason given in the last chapter where it was shown that to act for a reason involves an agent having a belief and a desire which gives the agent's goal. Because this belief and desire are of import to acting for a reason, and hence to the paradigm cases of acting rationally, the beliefs and desires of an agent can also be of import to whether or not other actions qualify as rational.

Before closing I will say how this fits with the use made of CNRs in Chap. 4. There I referred to two apparent CNRs: one ought if one believes one's mother unwell, to call around to see her; and, one ought if one desires to make a cake, to gather ingredients. The former of these is a specification of CNR1. The point at issue is that one's mother being unwell is a reason to call to see her, one believes that reason to obtain, and so one ought to have a goal giving desire to go around to see her. As such the best thing to say is that, if Si (the agent at issue) is mistaken in believing his mother unwell, he is irrational for not going around to see her. But, this does not mean that he ought to go around to see her. Rather, because she is well Si should carry on home to his children. The second CNR is a specification of CNR2. Gathering ingredients is a means to making a cake. An agent with the goal of making a cake is irrational if that agent does not gather ingredients despite believing that he or she must do so in order to achieve his or her goal. In the previous chapter the agent at issue is Dave. Dave desires to make a cake but his desire is ill founded and not had for a reason because he is mistaken in believing that it is his friend's birthday. We are not advising Dave what he ought to do in advising him that in order to make a cake he must gather ingredients. Rather we are evaluating certain actions as rationally excusable. Nevertheless, from Dave's perspective (unwilling to accept it is not his friend's birthday or resultantly to drop his desire) he will feel that he has reason to gather ingredients and hence the evaluation will strike him as advice on what he ought to do.

Notes

1. I will modify this claim shortly, but it remains true so long as we suppose I don't believe there to be any other reasons for me to eat or not eat the salmon.
2. This has led some to speak of subjective reasons (e.g. Cullity and Gaut (1997)), but I prefer to avoid such labels in order to avoid the idea that we are talking about a special type of reason.
3. Cf. Mele and Rawling (2003), Audi (2003), Smith (2003). However, some do go further by suggesting that the two also differ in form. E.g., Harman (2003) suggests that theoretical and practical reasoning differ in that the latter allows preferences to influence ones reasoning.
4. Scanlon (2007) offers the looser: an agent ought 'to r if that agent believes she/he has an object given reason to r'; or Broome (2007a) offers: an agent ought 'to r if she/he believes that she/he ought to r'.
5. Cf. Broome (2004).
6. At this stage I might also lay my cards on the table as far as seeing there is a reason is concerned. I do not take this to be a special sort of seeing as if one becomes aware of a fact and then one becomes aware of the fact having a special quality of being a reason. The reasons simply are facts and only need to be seen as such. The case of instrumental reasons is slightly more complex. E.g., I must be aware that I have a reason to walk in order to be aware that the fact that putting on shoes is a means to walking is a reason. But this does not involve seeing the fact that putting on my shoes is a means to walking in a new way or seeing something more about it. It is simply a matter of being aware of how this and other facts fit together. Cf. Foot (2001, ch. 4).

7. I am perhaps assuming that there will always be something that there is most reason to do. Korsgaard (2009) thinks that there is something awkward in believing this if one is a non-psychologist about reasons in the manner in which I am because it is not clear that the reasons must always point in one direction because it is not clear that facts must. But, I see little more problem in assuming that reasons thus conceived will always point in one direction than assuming, as Korsgaard does, that reasons must point in a single direction. Further, I will happily drop this assumption and allow that there might be cases in which two acts are equally reasonable.
8. Cf. Broome (2007a) and Korsgaard (1997).
9. Cf. Kolodny (2008), Broome (2007a), and Parfit (2001).
10. Cf. Broome (2007a).
11. Cf. Kolodny (2005, 557) who adds a negative form of CNR1: an agent ought not to *F* if that agent believes that he or she lacks sufficient reason to *F*.
12. Some even go so far as to suggest that all reasons derive in some way form requirements of rationality, e.g., Korsgaard (1997).
13. Cf. Dancy (2000) and Broome (1999).
14. Broome (2007b) does try to bring out a distinction in discussing an agent painting themselves into a corner such that if an agent does paint themselves into a corner, only a wide scope requirement can allow them a way out. But Kolodny (2007a) responds by saying that it is not clear we should allow the agent a way out. After all, this doesn't mean the agent is trapped. It merely means that the agent is irrational in an instance. Further, Korsgaard (2009) puts pressure in the other direction when she speaks of a wide scope cook book being rather useless. Her point being, and this really respects Kolodny's emphasis on process requirements, that wide scope requirements are poor guides: one could still follow the wide scope requirement deeper into the corner.
15. Broom thinks of rationality as given in this negative way with a focus on not breaking the CNRs. Cf. Broom (2007b) and Kolodny (2007a).
16. I do not mean to imply that the belief must be true by talking of its content. We could talk about the belief or desire itself instead of the content. This would not alter the argument. But I follow Kolodny in talking of content as I believe it is more plausible.
17. Cf. Kolodny (2007a) for a fuller account of this.
18. It is no better if we speak of the content of the lack of desire.

19. Again, it is no help to speak of the content of the lack of belief.
20. Cf. Kolodny (2007b, 251) and Raz (2005). Kolodny actually argues that RC2 must be replaced by a more complicated reasons claim. But I am not concerned to provide all the necessary details here. My intention is to show how the general picture works and so I stick with the more clear idea reflected in Raz's position.
21. Cf. Kolodny (2008, 371).
22. Kolodny actually raises some doubts about this idea and instead offers an alternative reasons claim (cf. Kolodny (2008)).
23. Kolodny has gone to lengths to provide more detailed accounts (cf. Kolodny (2005, 2007b, 2008)).
24. Cf. Raz (2005) and Kolodny (2005, 2007b, 2008). Scanlon says things which reflect aspects of both ideas (see Scanlon (2010) and Scanlon (2007) respectively).
25. In the case of CNR4 these are the beliefs that q and that q entails p, as the fact that q and the fact that q entails p are together reasons to believe that p.

References

Audi, R. (2003). Theoretical Rationality: Its Sources, Structure, and Scope. In A. Mele & P. Rawling (Eds.), *The Oxford Handbook of Rationality* (pp. 17–44). Oxford: Oxford University Press.

Broome, J. (1999). Normative Requirements. *Ratio, 12*, 398–419.

Broome, J. (2004). Reasons. In J. Wallace, M. Smith, S. Scheffler, & P. Pettit (Eds.), *Reason and Value: Themes from the Moral Philosophy of Joseph Raz* (pp. 28–55). Oxford: Oxford University Press.

Broome, J. (2007a). Does Rationality Consist in Responding Correctly to Reasons? *Journal of Moral Philosophy, 4*, 349–374.

Broome, J. (2007b). Wide or Narrow Scope? *Mind, 116*, 359–370.

Cullity, G., & Gaut, B. (1997). Introduction. In G. Cullity & B. Gaut (Eds.), *Ethics and Practical Reason* (pp. 1–27). Oxford: Oxford University Press.

Dancy, J. (2000). *Practical Reality*. Oxford: Oxford University Press.

Foot, P. (2001). *Natural Goodness*. Oxford: Oxford University Press.

Harman, G. (2003). Practical Aspects of Theoretical Reasoning. In A. Mele & P. Rawling (Eds.), *The Oxford Handbook of Rationality* (pp. 45–56). Oxford: Oxford University Press.

Kolodny, N. (2005). Why Be Rational? *Mind, 114*, 509–563.

Kolodny, N. (2007a). State or Process Requirements? *Mind, 116*, 371–385.

Kolodny, N. (2007b). How Does Coherence Matter? *Proceedings of the Aristotelian Society, 107*, 229–263.

Kolodny, N. (2008). The Myth of Practical Consistency. *European Journal of Philosophy, 16*, 366–402.

Korsgaard, C. M. (1997). The Normativity of Instrumental Reason. In G. Cullity & B. Gaut (Eds.), *Ethics and Practical Reason* (pp. 215–254). Oxford: Oxford University Press.

Korsgaard, C. M. (2009). The Activity of Reason. *Proceedings and Addresses of the American Philosophical Association, 83*, 23–43.

Mele, R., & Rawling, P. (2003). Introduction: Aspects of Rationality. In A. Mele & P. Rawling (Eds.), *The Oxford Handbook of Rationality* (pp. 3–14). Oxford: Oxford University Press.

Parfit, D. (2001). Rationality and Reasons. In D. Egonsson, B. Petterson, & T. Ronnow-Rasmussen (Eds.), *Exploring Practical Philosophy: From Action to Values* (pp. 17–39). Aldershot: Ashgate.

Raz, J. (2005). The Myth of Instrumental Rationality. *Journal of Ethics and Social Philosophy, 1*(1), (electronic publication).

Scanlon, T. (2007). Structural Irrationality. In G. Brennan, R. Goodin, F. Jackson, & M. Smith (Eds.), *Common Minds: Themes from the Philosophy of Philip Petit* (pp. 84–103). Oxford: Oxford University Press.

Scanlon, T. (2010). Metaphysics and Morals. In M. De Caro & D. Macarthur (Eds.), *Naturalism and Normativity* (pp. 173–192). New York: Columbia University Press.

Smith, M. (2003). Humean Rationality. In A. Mele & P. Rawling (Eds.), *The Oxford Handbook of Rationality* (pp. 75–92). Oxford: Oxford University Press.

Part III

The Reality of Tense and Emergence of the Self

7

The Argument from Rational Action

In this chapter I will take up and defend an argument mentioned in Chap. 3. There we saw that many actions will only be rational if their agents have tensed and/or first-personal beliefs. Mellor tried to account for this need in terms of the causal role of beliefs. However, I raised some doubts about the adequacy of this answer because it presupposed an asymmetry that did not in fact exist and it seemed to neglect the manner in which these actions are responsive to reasons, or rational. I took up this theme in the subsequent chapters. I argued that reasons, both in their role of justifying and motivating, are facts. I then argued that acting for a reason was a matter of being aware of that reason, and then forming a goal, choosing to act, and acting in response to that reason. I used this notion of acting for a reason to clarify the notion of rationality. A rational action, in the paradigm or non-error case, is an action which an agent does for reasons which provide all things considered reason for that action. In an error case, when an agent mistakenly takes a reason to obtain which does not obtain, then that agent may qualify as rational on the basis of the similarity between that case and a paradigm case.

I will bring these elements together in this chapter to argue that if rationality demands tensed and first-personal beliefs in place of tenseless and non-first-personal beliefs, then this must be because the former

Olley (F.O.C.H.) Pearson, *Rationality, Time, and Self*,
https://doi.org/10.1007/978-3-319-71973-3_7

involve an awareness of facts that the latter does not. Both tenseless and non-first-personal beliefs can be involved in acting for reasons. What distinguishes them from tensed and first-personal beliefs in the eyes of rationality is simply the reasons they are an awareness of. It follows from this that the tenseless theory of time is wrong because it denies that tensed beliefs capture facts not captured by tenseless beliefs. In addition, we are moved towards an emergentist theory of the self according to which it is something over and above the body which can be readily captured by third-personal language. This does require a realist and pluralist ontology which I will also show to be independently plausible.

7.1 The Argument from Action

My first argument against the tenseless theory of time and towards the emergence of the self concerns rational actions and runs as follows:

P1. In some cases an agent must have a tensed belief in order to act rationally and no tenseless belief can satisfy this requirement.

P2. In some cases an agent must have a first-personal belief in order to act rationally and no non-first-personal belief can fulfil this requirement.

P3. The reasons that motivate actions (m-reasons) and the reasons that justify actions (j-reasons) are facts.

P4. In non-error cases of rational action if an agent is required to have a belief and another belief cannot satisfy that requirement, then the former must be an awareness of a j-reason that the latter is not.

C1/P5. If there are non-error cases of rational action in which an agent is required to have a tensed belief and a tenseless belief cannot satisfy that requirement, then the former must be an awareness of a j-reason that the latter is not. (From P4.)

C2/P6. If there are non-error cases of rational action in which an agent is required to have a first-personal belief and a non-first-personal belief cannot satisfy that requirement, then the former must be an awareness of a j-reason that the latter is not. (From P4.)

C3/P7. If there are non-error cases of rational action in which an agent is required to have a tensed belief and a tenseless belief cannot satisfy that requirement, then the former must be an awareness of a fact that the latter is not. (From P3 & P5.)

C4/P8. If there are non-error cases of rational action in which an agent is required to have a first-personal belief and a non-first-personal belief cannot satisfy that requirement, then the former must be an awareness of a fact that the latter is not. (From P3 & P6.)

P9. There are non-error cases of rational action in which an agent is required to have a tensed belief and no tenseless belief can fulfil that requirement.

P10. There are non-error cases of rational action in which an agent is required to have a first-personal belief and no non-first-personal belief can fulfil that requirement.

C5/P11. Agents are aware of facts in tensed beliefs that they are not aware of in any tenseless beliefs. (From P7 & P9.)

C6. Agents are aware of facts in first-personal beliefs that they are not aware of in any non-first-personal beliefs. (From P8 & P10.)

P12. According to the tenseless theory of time there are no facts captured by tensed beliefs that cannot be captured by tenseless ones.

C7. The tenseless theory of time is wrong. (From P11 & P12.)

I shall refer to this as *the argument from action*. In this chapter I will focus on the temporal case. It is slightly less clear what conclusions we should draw about the self and I will say more about these in Chap. 10.[1]

I will now set about justifying the premises that I take to need external support, namely P1-4, P9, P10 and P12. The other premises gain their support from one or more of these.

P12 comes straight forwardly from the definition of the tenseless theory that was given in Chap. 2 and the manner in which beliefs are characterized. A tensed belief will be a belief an agent has when that agent is disposed to honestly express or affirm uses of tensed language. It will be a belief that an agent communicates by using tensed language. It is therefore natural to take the beliefs and the language to share contents and to capture the same facts. Therefore, if tensed beliefs capture facts that tenseless ones do not, then tensed language captures facts that tenseless language does not. The definition of the tenseless theory is that it denies that tensed language captures facts that tenseless language does not. Therefore, we arrive at P12.

P1 and P2 come from Chap. 3. There we saw that Perry's argument for the essential indexical has showed that some actions require their agent to have tensed beliefs. A consideration of this and Mellor's arguments led to the conclusion that the tensed beliefs were essential specifically for some rational actions. If an agent is going to act rationally and in a timely fashion, then that agent must have a tensed belief. For instance, I must believe that 'it is now 1pm' or something similar and tensed if I am going to act rationally to fulfil my goal of listening to the 1 o'clock news. I might have tenseless beliefs such as 'listening to the news requires the radio to be on' and 'in order for the radio to be on at 1 o'clock it must be turned on before 1 o'clock' all morning, but in order to act rationally in turning on the radio I must also gain a tensed belief. No tenseless belief can fulfil this role. Therefore, P1 is correct.

Chapter 3 also showed that analogous remarks apply for a first-personal indexicals. An agent must have a first-personal belief if that agent is going to act rationally and for considerations that apply to him or her in particular. For instance, if I am going to collect my order I must believe 'I have ticket 114' or something similar and first-personal. No non-first-personal belief can fulfil this role. For example, the belief 'Olley has ticket 114' will not suffice because I might not know that I am Olley. Therefore, P2 is correct.

The discussion of Chap. 4 showed that reasons are facts. More specifically, it was shown that the reasons that make an action appropriate or justified are facts, and that the reasons that an agent acts for are facts. It is the fact that cycling gives me pleasure that justifies and motivates me to go cycling on a particular occasion. This gives us P3.

7.2 Justification of P9 & P10

P9 and P10 say that there are non-error cases of rational action which require tensed and first-personal beliefs. My reason for taking these premises to be correct is that denying them leaves one in an unacceptable position.

A great majority of our actions are timely actions and/or actions done for reasons specific to ourselves. That is, a great many of our actions fall within the scope of Perry's essential indexical arguments. (I have argued in Chap. 3 that all of Perry's examples—so those concerning 'here', 'this', or 'you' beliefs too—can be interpreted as concerning tensed and/or first-personal beliefs, and few of our actions escape all of his arguments.) To deny P9 and P10 would thus entail that a great majority of our actions are error cases.

Because reasons are facts saying that the great majority of our actions are error cases implies that they are cases in which we are mistaken about what facts obtain. It would follow that people were generally mistaken in their world view. But, this is something that we should deny and that the tenseless theory has been careful to deny. In its old form this denial occurred in the idea that tensed beliefs are reducible to tenseless ones. In its new form it occurs in the idea that tensed beliefs have tenseless truth-conditions. Either way we are encouraged to think that our tensed characterizations get the world right.

If the tenseless theorist tried to avoid my argument by insisting that most of our actions are error cases, then this means the contents of our tensed and first-personal beliefs are mistaken. (Essential indexical cases can be constructed using all sorts of tensed or first-personal beliefs.) If, as it seems, the content of our perceptual beliefs are all present tensed,[2] it follows that most or all of our perceptual beliefs are mistaken. Our perceptual beliefs get the world wrong. This mistake would thus infect all of our empirical data and hence even the sciences—commonly revered by tenseless theorists—would be shown to be built on error. Whilst I am ready to admit that we do not get everything right, error on this scale is surely unacceptable.

Moreover, the tensed and/or first-personal beliefs are required for the actions at issue to be rational actions. This means that the reasons that they are an awareness of, or would be if these were not error cases, are reasons that are essential for the actions. They are reasons required to justify the actions. It follows that if all of these actions are error cases, then they are without reasons that would be essential for justifying them. In other words, essential justificatory reasons would fail to obtain for any of these actions. The actions would hence all be unreasonable actions. However, this sounds wrong. Furthermore, it also implies that we ought to be acting quite differently to the ways we do act. But, it is quite incomprehensible to fathom what such a radical change in people's actions ought to be. If I shouldn't turn the radio on at 1pm, or go to the meeting at 2pm, or buy milk when I run out, or any of these mundane everyday actions, I am lost to imagine what I ought to do. (One could not avoid these consequences by denying that the actions at issue were rational actions, because, as I argued in Chap. 4, it is not possible for the majority of our actions to be irrational.[3])

In short, I take it that P9 and P10 must be correct because to deny them is to insist that the great majority of our actions are unreasonable. Further, it would be to suggest that the great majority of our world view is mistaken. I do not think that these consequences are acceptable and hence P9 and P10 must be.

7.3 Justification of P4

Beliefs entered our discussion of rational action on the following basis. To act rationally an agent must act for a reason that justifies that action. (This might need to be altered to accommodate error cases but there is no need to add this complexity here.) To act for a reason an agent must be aware of it. Beliefs provide awareness. Therefore, beliefs are required for rational action.

This reveals the role beliefs have in our consideration of the rationality of actions. By considering this role that beliefs play we can uncover what their rational import is. More specifically, we can see that there are three elements to their rational import. First, beliefs must provide awareness. Second, the awareness provided must be of justifying reasons. Third, beliefs must enable an agent to act for those reasons.

It follows that if two beliefs are of differing rational import they must differ in one or more of the following three ways. First, one must be a state of awareness whilst the other is not, RIB1. Second, one must be a state of awareness of a justifying reason whilst the other is not, RIB2. Third, one must enable the agent to act for a reason whilst the other does not, RIB3.

P4 simply states that RIB2 must occur when two beliefs are of differing rational import. Therefore, one way to defend P4 is to rule out RIB1 and RIB3. RIB1 is not actually an option at all. According to the umbrella notion of belief that has been assumed throughout this book beliefs simply are states of awareness. In the rest of this section I will thus argue against RIB3. In short, I will argue that tensed and tenseless beliefs alike enable an agent to act for a reason, and first-personal and non-first-personal beliefs alike enable an agent to act for a reason.

We saw in Chaps. 5 and 6 that a paradigm case of rational action will involve an agent being aware of a reason for an action, and then choosing to perform that action for that reason, adopting the goal of performing that action, and acting. We thus have four components of acting for a reason: belief, choice, desire, and action. The question of whether a belief can enable an agent to act for a reason is hence the question of whether it can combine with choice, desire, and action in the right way. Accepting RIB3 would involve arguing that some beliefs cannot combine with choice, desire, and action in the right way. However, this is highly implausible. It would mean that an agent could be aware of a reason and yet for it to be impossible for that agent to act for that reason.

An agent can fail to recognize that he or she is aware of the same thing on two different occasions. One might hence wonder if this is enough to establish RIB3: in one state of awareness an agent is aware of a reason and is enabled to act for it; however, in the other state of awareness the agent fails to recognize that he or she is aware of the same thing or hence to act. But, this argument is no good. Crucially, an agent is not required to recognize that he or she has been aware of a reason on another occasion in order to be able to act for that reason. Furthermore, even if we supposed such recognition was necessary it does not establish RIB3 or deny P4. Firstly, while it is possible to fail to recognize that one is aware of the same thing on two different occasions it is not necessary to do so. But, it would have to be necessary if it were to ground the impossibility of acting for reasons through specific beliefs. Secondly, such cases of misrecognition appear to be possible because different aspects of that thing are presented on each occasion. For example, I might not recognize that I am looking at the same house on two occasions because on one I was aware of the front aspect and on the other I was aware of the rear aspect. But, being presented with different aspects in this way just is being presented with different facts. For example, being close to the rear of a house is a different fact from being close to the front of a house. Therefore, we are left with the idea that the different beliefs provide awareness of different facts and hence we are led back to P4 and away from the idea that possible failures of recognition can justify RIB3.[4]

It appears that an agent's beliefs can vary independently of their choices, desires, and actions, contra RIB3. For example, I can believe that walking gives me pleasure, and can choose not to walk, desire not to walk, and not go walking for this reason (perhaps I feel in need of some self-discipline). Alternatively I might choose to go walking, desire to go walking, and go walking for this reason. Moreover, I might believe that walking gives me pleasure and choose to cycle, desire to cycle, and go cycling. That is, I can choose to go cycling because walking gives me pleasure. We can make this case appear plausible by adding in some other beliefs to the effect that people who gain pleasure from walking gain pleasure from cycling. However, there is no need to do this. If we do not, then I will be acting irrationally. The fact that one gains pleasure from walking is not a j-reason to go cycling. This makes it hard to understand such a case because we generally try to understand people by assuming them to be rational. But, this does not show that the case is impossible. The requirement that we take other agents to be rational in order to understand them is a proportional requirement. We must take another to be rational in a good proportion of cases, but we do not need to take that person to be rational in every case. It is clear we can understand others well enough to pick out their occasional irrational actions as irrational on the basis of a background of their rational actions. In fact, there are examples in psychology of apparent widespread irrationality of this sort. It is common for people to perform actions for reasons that are not reasons to do those things or are not reasons that justify those actions.[5]

Acting for a reason or a fact one takes to be a reason has a certain phenomenology. Agents can normally tell you when they are acting for reasons and what those reasons are. In the terminology of Chap. 4, one might say having an m-reason has a certain phenomenology whether or not that m-reason is also a j-reason. This can enable one to know when another is acting for a reason and hence when he or she is acting irrationally. If I cannot make sense of another's actions I might ask him what his reasons for performing those actions were. It might turn out that the facts he refers to are not j-reasons for those actions at all and I might hence judge him to be irrational. But, crucially, it is not necessary that I conclude that he did not in fact decide or desire to act. It is clear people are sometimes wrong about what a particular fact justifies or what it is a j-reason for. If we can talk about being more or less off target in this respect, then each of these are as possible as the other even if they are not equally common. Irrational actions show just how freely beliefs can vary independently of choices, desires, and actions in a way that conflicts with RIB3. All manner of beliefs are capable of both combining with or not combining with desires, choices, and actions. Such combinations might not be expected, but they are possible.

It can be appropriate to take an agent's movements to be simply non-rational so that it is inappropriate to judge that agent in terms of rationality at all. Breathing is an example of a non-rational movement. Being manipulated by someone else is a different example. However, these cases are clearly different to the ones I just described as irrational. In the cases of irrational action the agents have their action as a goal and they choose to perform that action on the basis of a fact they are aware of (which as it happens is no j-reason for the action that they perform). These conditions are lacking in the examples of non-rational actions I outlined and it appears to be this very lack which renders them non-rational.

When considering the argument from rational action our focus is on tensed and first-personal beliefs. The relevant specifications of RIB3 are thus as follows. First, a tensed belief enables acting for a reason whilst a tenseless belief does not. Second, a first-personal belief enables acting for a reason whilst a non-first-personal one does not. But, both of these claims seem simply to be false. Perry's arguments for the essential indexicals utilize specific sorts of cases. For example, we need a tensed belief when we must act at a specific time. It thus appears that there will be cases of rational actions in which an agent does not need a tensed belief. These might be cases in which an agent does not need to act at a specific time or in which the agent has a reason to do something at any time at all. If this is right, it shows that tensed beliefs are not required simply in order for a person to act rationally. It shows that beliefs other than tensed ones can combine with choices, desires, and actions in the relevant way in rational cases of acting for a reason. Similarly, the need for first-personal beliefs appears to arise in cases where it is oneself in particular who has a reason to act. So if there are cases in which this is not so, cases in which anyone or everyone ought to act, then these will be cases where first-personal beliefs are not required for rational actions or acting for a reason. These will be cases in which non-first-personal beliefs combine with desires, beliefs and actions in the right way. Moreover, it is common for tenseless and non-first-personal beliefs to play roles in many actions. For example, when I am discussing mathematics with a friend I might well make use of a tenseless non-first-personal belief such as 'four plus four is equal to eight'. Tenseless and non-first-personal beliefs readily combine with choices, desires and actions in an agent's acting for a reason.

It is worth remembering that in thinking that an action is done for a reason we are not thinking that the action is caused by the reason or caused in a specific way.[6] It follows that it is inappropriate to say that only beliefs with a particular causal shape can combine with choices, desires, and actions in the right way to enable an agent to act for a reason. Rather, if we are able to say that an agent desires, chooses and acts on the basis of what they believe, then we can also say that if there are any causal requirements, then these must also have been met.

RIB3 is false. Tenseless and non-first-personal beliefs can enable agents to act for reasons just as tensed and first-personal beliefs can. This becomes particularly clear when we consider irrational actions or actions outside the scope of Perry's arguments. It follows that if tensed beliefs are required for a rational action in place of any tenseless beliefs, then the former must involve an awareness of reasons that the latter lacks. Similarly, if first-personal beliefs are required for rational actions in place of any non-first-personal beliefs then the former must involve an awareness of reasons that the latter lacks. P4 is correct.

7.4 Facts

So far I have spoken a great deal of facts. Facts are reasons and are where the reality of tense lies. However, I have said very little about what exactly a fact is. In this section I will say something to address this. I will do this by looking back at my arguments and discussions and what they presuppose facts to be like. I have used 'fact' in a loose everyday way. It will turn out that my use can be taken to coincide with the idea that facts are an ontologically fundamental kind, but my use is not committed to such an ontology. My facts could simply be, for example, particulars instantiating modes, bundles of tropes containing specific tropes, or, particulars standing in specific resemblance relations.[7]

Three aspects of my use of facts stand out. The primary role facts have played so far is in providing a form of realism or objectivity. Facts are aspects of the world. They are what we consider in acting and are in contrast to the intentional and intensional states through which we make these considerations. Secondly, my arguments presuppose or entail that there is a plurality of facts. The fact that I like cycling is different from the fact that my knee is injured and it is because they are different that they can provide different reasons. (The former is a reason to go cycling and the latter a reason not to.) Thirdly, it appears from my discussion that facts can be general, negative, disjunctive, and conjunctive. For example, it appears that one of my reasons for going to the shop might be the fact that there is no milk anywhere in my house. I will argue all three of these aspects are unproblematic.

My view of facts is realist in two senses. Firstly, it supposes that facts exist. Secondly, it supposes that facts are objective, or mind independent. I assume that both of these claims are acceptable and I will say little to defend them from sceptical attacks. I should, however, clarify the objectivity claim. When I say that facts are mind independent I do not mean that all the facts would remain the same if there were no minds, because some facts plausibly involve the existence of minds. For example, the fact that I feel happy involves my mind. However, these mind involving facts remain objective insofar as no one need recognize or consider them for them to obtain. This objectivity was brought to the fore by my examination of reasons where it was shown that reasons, insofar as they are as facts, are starkly contrasted with beliefs which are the presentations of these facts. This last comment makes clear that I also assume that facts can be objects of awareness. We can come to know a fact obtains. Again, I assume this is plausible and will not address sceptical concerns to the contrary.

Armstrong[8] has defended an ontology according to which states of affairs are ontologically fundamental and therefore real and objective. A state of affairs is a complex constituted by a particular object and a property, or *n* objects and an *n*-ary relation. These properties and relations are considered to be universals. They are repeatable and identical in each repetition. Armstrong's states of affairs appear to coincide nicely with my facts. I recognize it to be a fact that the road is slippery and Armstrong would in the same way take the road's being slippery to be a state of affairs. It would be a complex of the particular road and the property of slipperiness.[9] With this in mind I will show how my notion of a fact is plausible by assimilating it to Armstrong's notion of a state of affairs.

If one is going to adopt a states of affairs ontology, one must say something about how a state of affairs is composed: how should a particular and a property be related when the particular instantiates the property? The state of affairs of a ball being red does not simply require the existence of the ball and the property of redness because it is clear that both the ball and redness can exist and the ball not be red. For example, if the ball is yellow and a bat is red. Thus, one must say that the property and the particular are related in some way. However, difficulties arise when we consider this relation. Let us call it 'R'. For example, it is not sufficient to simply take R to be a third ingredient in the state of affairs alongside the ball and the property of redness, because it appears that the ball, R, and the property redness can all exist when the ball is not red. Plausibly, all three exist when the ball is yellow and the bat is red. Redness exists in the bat. The ball exists. R exists as the relation between the ball and yellowness. If one tries to deal with this problem by postulating a further relation, R2, to hold between R and the ball and redness, then clearly the same issues simply arise again. This problem will always reappear if we simply postulate more relations like R. Such an infinity of relations appears to be unacceptable and to this extent the state of affairs of the ball being red is thrown into doubt. (This is a variant of Bradley's regress.[10])

However, Armstrong's theory can avoid this difficulty because for Armstrong[11] the relation between the ball and the property redness is a relation that could not occur without relating those two. Therefore, we do not need to postulate a further ingredient to relate this relation, the ball, and redness. The same applies to other states of affairs. The constituents of the state of affairs are internally related. Armstrong says this about the relation because he takes the state of affairs of the ball being red to be ontologically prior to the ball or the property of redness, because the latter are abstractions from the former (though the ball can/will possess other properties and other particulars can/will be red).[12]

A second problem for a states of affairs ontology directly relates to my assumption that there is a plurality of facts. It has been argued that a states of affairs ontology is implausible because it entails that all states of affairs are identical. The argument I have in mind has become known as the slingshot argument and can be presented as follows:[13]

P0. Jill is sad and Henry is happy.
P1. The state of affairs that Henry is happy is identical to the state of affairs that Henry is happy.
P2. A statement such as occurs in P1, if true, cannot be falsified by replacing the embedded sentences (for example, Henry is happy) by logically equivalent sentences.
P3. A statement such as occurs in P1 cannot be falsified by replacing a referring term in it by a co-referring term.
P4. The sentence 'Henry is happy' is logically equivalent to the sentence 'the unique *x* such that (*x* is identical with Jill) is identical to the unique *x* such that (*x* is identical to Jill and Henry is happy)'. (Given P0.)
P5. The sentence 'Jill is sad' is logically equivalent to the sentence 'the unique *x* such that (*x* is identical with Jill) is identical to the unique *x* such that (*x* is identical to Jill and Jill is sad)'. (Given P0.)
P6/C1. The state of affairs that Henry is happy is identical to the state of affairs that the unique *x* such that (*x* is identical with Jill) is identical to the unique *x* such that (*x* is identical to Jill and Henry is happy). (From P1, P2 & P4.)
P7/C2. The unique *x* such that (*x* is identical to Jill and Henry is happy) is identical to the unique *x* such that (*x* is identical to Jill and Jill is sad). (From P4 & P5.)
P8/C3. The state of affairs that Henry is happy is identical to the state of affairs that the unique *x* such that (*x* is identical with Jill) is identical to the unique *x* such that (*x* is identical to Jill and Jill is sad). (From P3, P6 & P7.)
C4. The state of affairs that Henry is happy is identical to the state of affairs that Jill is sad. (From P2, P5, & P8.)

P0 simply assumed any two true sentences and therefore the sentence 'Jill is sad' could be exchanged for any other true sentence. This means that the argument form can be used to show that any state of affairs is identical to the state of affairs that Henry is happy. Likewise, one could swap the sentence 'Henry is happy' for any other true sentence. Therefore, we could argue that any two states of affairs are identical.

One way of responding to the slingshot argument is through denying the move from C2 to C3. One can do this if one adopts Russell's theory of descriptions according to which definite descriptions do not refer.[14] If this is right, it follows that the definite descriptions *the unique* x *such that (*x *is identical to Jill and Henry is happy)* and *the unique* x *such that (*x *is identical to Jill and Jill is sad)* do not refer. If they do not refer, then P3 does not apply to them and we cannot get to C3.[15] An alternative way to deal with the slingshot is to deny either P2 or P3. Mellor (1999, ch. 9.5) denies P3, though he is careful to point out that this doesn't entail the falsity of the plausible idea that a statement such as occurs in P1 cannot be falsified by replacing a term in it that refers to a fact by a co-referring term, P3*. However, if P3 were replaced by P3*, the slingshot argument would not work. Searle[16] instead denies P2. The power of these responses lies in the fact that P2 and P3 are difficult to justify whilst the conclusion C4 appears to be obviously false. It is more plausible to deny C4 and deny P2 or P3, than to accept all three.

Armstrong (1997, ch. 8.6) offers two types of identity criteria for states of affairs. Firstly, he says that different states of affairs can be distinguished by their constituents. So, states of affairs that are constituted by different particulars or different universals will be different states of affairs. Secondly, Armstrong accepts that we can often distinguish between states of affairs empirically. In particular, two states of affairs which are not necessarily concurrent are not identical. For example, the state of affairs of Bill loving Bess has the same constituents as the state of affairs of Bess loving Bill. However, unfortunately, it is possible for one of these states of affairs to occur whilst the other does not and therefore it is clear that the two are distinct states of affairs.[17] This last example also highlights that for Armstrong a state of affairs cannot merely be identified by its constituents. It must be identified in part by the manner in which these constituents are structured. Bill loving Bess forms a different structure from Bess loving Bill.

Thus far Armstrong's states of affairs are adequate to play the role of my facts. They are realist and there is a plurality of them. This raises the question of how Armstrong deals with negative, general, disjunctive, and conjunctive states of affairs, and whether this is adequate to my concerns.

The conjunctive case is perhaps the easiest to deal with. Armstrong (1997, 134, 2004) denies that there are conjunctive states of affairs. The state of affairs of a ball being red and round is nothing over and above the state of affairs of the ball being red and the state of affairs of the ball being round. This is perfectly acceptable for my notion of fact which does not require a conjunction to be anything over and above the conjuncts. Plausibly one is not rationally required to do something because of *x* and *y* rather than because of *x* and because of *y* (or because one believes *x* and *y* rather than because one believes *x* and believes *y*).

Armstrong (1997, 134, 2004) also denies that there are disjunctive states of affairs. Here it is perhaps useful to note his use of the truth-maker relation. He says that a statement such as 'the ball is red or it is round' is made true by either the state of affairs of the ball being red, or the state of affairs of it being round, or by both. Because of this he believes that no special disjunctive fact is needed as a truth-maker. This might raise trouble for my use of facts if I am to say that a rational action might require an agent to have a disjunctive belief in place of a belief in one or the other or both disjuncts. Because, such a case would appear to show that the disjunctive belief must get at a fact that the combination of disjuncts cannot. This would perhaps be a disjunctive fact. However, such a potentially problematic case is not possible. If it is rational for me to believe 'either Jack or Bob are at the gallery' and therefore to form a goal of going there, then it is equally rational for me to believe simply 'Jack is at the gallery' and therefore form the goal of going, or to believe 'Bob is at the gallery' and therefore form the goal of going, or to believe 'Jack and Bob are at the gallery' and therefore form the goal of going. This is because the disjunction can only be a reason if each of or at least one of the disjuncts alone would be a reason. As non-disjunctive beliefs can step in for disjunctive ones, we needn't suppose that disjunctive beliefs get at facts that the non-disjunctive ones do not, nor hence posit disjunctive facts.

One might worry that there could be a case in which it is rational to act on a disjunctive belief, but not on either disjunct, if one knows one or the other disjuncts to obtain but not which one obtains.[18] However, this simply puts matters back a step. We must consider the evidence for the disjunction. If this evidence is not disjunctive, we can simply refer to this non-disjunctive evidence in place of the disjunction. Plausibly our evidence is non-disjunctive at root because we do not see or experience disjunctions. So, we can always step back from a disjunctive belief to non-disjunctive beliefs. For example, I might believe that 'one of my friends is at the gallery' and also believe that 'I have two friends, Bob and Jack'. On the basis of the evidence these beliefs provide I could believe 'either Bob or Jack are at the gallery' and therefore form the goal of going there. But, plausibly, this disjunctive belief gives me a reason to form that goal only because the non-disjunctive beliefs on which it was based gave me a reason. I have a reason to go there because one of my friends is there. So, it is plausible that rationality will never require an agent to have a disjunctive belief in place of a non-disjunctive one. Any act that would be rational if the agent had a disjunctive belief would be equally rational if the agent instead had one or more non-disjunctive beliefs either concerning one or more of the disjuncts or evidence for the disjunction. Therefore, we do not need to suppose that disjunctive beliefs get at facts that non-disjunctive ones do not or that there are disjunctive facts.

We can distinguish two types of general beliefs. For example, I might believe 'some apples are red', or 'all apples are red'. It is easiest to treat these cases separately.

Armstrong does believe in general states of affairs: he believes in totality states of affairs. Suppose that there are only two apples. The state of affairs that all the apples are red will be a state of affairs over and above the state of affairs of the first apple, a1, being red and the state of affairs of the second apple, a2, being red. Armstrong supports this idea by noting that these two apples could exist and be red, and there also exist a further green apple. In this latter case there would not be a state of affairs of all the apples being red even though there remained the state of affairs of a1 being red and the state of affairs of a2 being red.

Armstrong speaks of a totalling relation when considering these general facts. In our example a totalling relation binds all the apples. Specifically, the totalling relation holds between the mereological whole composed of all the apples taken as a particular and the property *being an apple*.[19] Thus, the truth-maker for 'all apples are red' will involve a1 being red, a2 being red, and the totality state of affairs that these are all the apples. We can say that the belief that all the apples are red gets at a different state of affairs to the belief that a1 and a2 are red. Armstrong believes in totality states of affairs and so I can readily admit that totality beliefs, beliefs concerning all the so-and-sos as all of the so-and-sos, capture different facts from non-totalling beliefs. Therefore, I can freely admit that rationality requires totalling beliefs in place of beliefs concerning each and every individual in the total. (Note, I can admit them. I am not suggesting that I must.)

For Armstrong the belief that some apples are red will be made true by any state of affairs of an apple being red. For example, if a1 is red, this state of affairs will make it true that some apples are red even if a2 is green and a1 and a2 are all the apples. To this extent Armstrong does not think there is a general state of affairs of some apples being red. We thus arrive at a potential difficulty for my position. If rationality can require an agent to have a some-belief (e.g. 'some apples are red') in place of any particular belief (e.g. 'a1 is red'), then this would imply that the some-belief captures a fact not captured by any of the specific beliefs. This might be a special some-fact. If I must admit that there are some-facts and Armstrong does not admit that there are some-states of affairs, then my facts cannot be identified with his states of affairs and I cannot simply adopt his ontology. The problem is similar to the disjunctive case and so too is the solution. Such potentially problematic cases cannot arise because if it were the case that some apple being red was a reason to act, then a specific apple being red would also provide a reason. So, it will not be the case that rationality requires a some-belief rather than a particular belief that would fall under this general one. For example, suppose I have reason to collect something red and I know that some apples are red and all the apples are in my room, then I could rationally go to my room. However, I could equally rationally go to my room if I know that a1 is red rather than merely that some apple is red. Rationality will not demand a some-belief in place of any specific belief, therefore it does not need to be the case that there are some-facts over and above any specific facts.

This brings us finally to the case of negative beliefs and hence negative facts. To a large extent Armstrong denies the existence of negative states of affair (1997, ch. 8.7, 2004, ch. 5). However, he believes that the totality states of affairs alluded to above are a form of negative state of affairs because they involve an element of 'and nothing more'. Armstrong believes that with totality states of affairs we have all the states of affairs that are needed to provide truth-makers for negative statements. For example, the statement that there are no green apples will be made true by the state of affairs of a1 being red, the state of affairs of a2 being red, and the totality state of affairs that those are all the apples. At first blush it thus seems that Armstrong can have a unique state of affairs to provide a truth-maker for a negative belief without having a (strictly) negative state of affairs. However, this does not offer me a ready solution for dealing with a rational requirement for negative beliefs. The problem here becomes apparent when we think of the truth-maker that Armstrong might offer for the belief that there are no yellow apples. Presumably, Armstrong would give this belief the very same truth-maker as he gave the belief that there are no green apples. This would be the states of affairs of a1 being red and a2 being red, and the totality state of affairs that a1 and a2 are all the apples. But, it is plausible that rationality can demand an agent to believe 'there are no red apples' instead of 'there are no yellow apples' and hence that they ought to capture different facts. Thus, my facts and Armstrong's states of affairs appear to part company. However, there is more of relevance to the beliefs than these truth-makers. For example, the fact that there are no green apples could also be made true by a set of states of affairs including the totality of the green objects and the states of affairs that none of them are apples (e.g. the state of affairs that the first green object is a car, the state of affairs that the second green object is a tree…etc.).[20] These states of affairs will not be of relevance to the belief that no apples are yellow. But, the belief that no apples are yellow will be related to a different set of states of affairs. It will be relevant to the totality state of affairs of all the yellow objects and the states of affairs of those things not being apples. With these extra states of affairs in mind we do appear to have enough positive states of affairs to provide the reasons captured by negative beliefs that agents rationally act on. The belief that no apples are red will be linked to states of affairs that the belief that no apples are yellow is not linked to and vice versa. Nonetheless, this suggestion also relies on it being plausible that rationality will not demand an agent to have a negative belief in place of a positive belief which captures the truth-makers just discussed. That is, it cannot be the case that rationality demands an agent to believe 'there are no yellow apples' in place of believing 'a1 is red, a2 is red and a1 and a2 are all the apples', or 'y1 is a banana, y2 is the sun, and y1 and y2 are all the yellow things'. If rationality were to make such a demand this would imply, via the argument from action, that the former belief captured a fact not captured by the latter two. However, it does not seem that there is any such demand. The positive states of affairs appear to provide all the reasons of the negative ones. It therefore appears that I needn't posit any negative facts other than totality facts and that the states of affairs that Armstrong considers as truth-makers for negative beliefs are plausibly all the states of affairs that are required to provide reasons in the cases at hand.

In short, Armstrong takes states of affairs to be real, objective, and to form a plurality. He argues that there are totality states of affairs but otherwise there are not general, negative, disjunctive or conjunctive states of affairs. Armstrong supports this view on the grounds that the states of affairs he posits are adequate to provide truth-makers for all truths. My facts can be identified with Armstrong's states of affairs because these are also sufficient to provide all the reasons one might need for rational actions. If I identify my facts with Armstrong's states of affairs I can also avoid the problems of Bradley's regress and the slingshot argument. Further, my facts can gain the support that Armstrong's states of affairs have because they provide truth-makers for all truths.

Nonetheless, this notion of fact could doubtlessly also be accommodated into a variety of other ontological systems. Crucially, a number of ontologies can be realist and objective, and can distinguish a plurality of facts. This is true of pure trope theory, a nominalist ontology that adopts only particulars (not properties), and a multi-category ontology. Further, Armstrong is not the only ontologist to posit fundamental entities of a slightly negative kind. For example, Russell at times posits negative facts and general facts, and Martin posits absences.[21] Therefore, if I need to posit totality or slightly negative facts, then my facts could plausibly be fitted to either Russell's or Martin's ontologies. Moreover, there is nothing essential to a pure trope, resemblance nominalist, or multi-category ontology that prevents them from adopting something analogous to Armstrong's totality state of affairs and the need for truth-makers for negative truths provides an independent reason for positing them. However, my concern here is merely to show that my notion of fact can be accommodated within an ontology and Armstrong's view establishes this. I am not concerned to argue which ontology best fits my notion of fact.

7.5 Summary

In this chapter I have argued that the tenseless theory of time is wrong. I argued this by making use of the conclusions of the previous chapters. Tensed beliefs are required for rational actions and they cannot be replaced by tenseless beliefs in this role. However, the role of beliefs in rational actions is simply to provide an awareness of reasons that are facts. It follows that tensed and tenseless beliefs must involve an awareness of different facts. The tenseless theory is wrong. This is the argument from action. The argument from action also showed that first-personal beliefs capture facts not captured by any non-first-personal beliefs. I will consider this conclusion in Chap. 10.

The argument from action makes clear what is demanded of my theory of facts. It must be realist, it must allow for there to be a plurality of facts, and, perhaps, it must recognize the existence of totality facts. All of these demands are met by identifying my facts with Armstrong's states of affairs. Though, they could plausibly also be met by other ontologies.

Notes

1. I provide a related argument for a related but weaker conclusion in (Pearson 2015).
2. Cf. Le Poidevin (2007, 78).

3. Cf. Davidson (1973/2001), Kim (1988, 392–393), and Wittgenstein (1953/2001b, pt. I, sect. 242).
4. I am not sure it is right to talk of the mode of presentation of a proposition, but if it is this is of no relevance because it is the propositional object that counts here. The object of the proposition is the reason. Therefore, so long as one can grasp the object of a proposition—that is, grasp a fact at all in a belief—then this is all that is of import to acting rationally. (Writers who discuss the mode of presentation of a proposition include Kaplan (1989), Salmon (2006), Braun (2006), Soames (2002) as discussed by Schiffer (2006, 2003, ch. 1).)
5. Wason (1966) provides a well-known example of this.
6. Cf. Chap. 5.
7. Cf, respectively, Lowe (2006), Campbell (1990), and Rodriguez-Pereyra (2002).
8. Cf. Armstrong (1997). Armstrong is taking his lead from Wittgenstein (1921/2001a) and Russell (1968). Skyrms (1981) follows this lead and adopts an ontology of states of affairs or facts (in the technical sense, rather than in my open sense), but his position differs from Armstrong's. Simons (1992) again follows this lead, but does not adopt an ontology of facts.
9. Armstrong certainly speaks in this manner, though we should be careful here because Armstrong does not want to be committed to objects or properties being of a specific form. It could, e.g., turn out that the only real objects are the fundamental particles of physics and the only real properties and relations are the properties and relations that these objects instantiate. However, this does not need to entail that there are not macroscopic things or qualities, or macroscopic states of affairs. It merely requires that these are strictly speaking complexes of the more fundamental states of affairs. Therefore, this does not affect my argument.
10. Cf. Bradley (1897, bk. I, ch. II).
11. Cf. Armstrong (1997, ch. 8) especially p. 118.
12. This leads Lowe (2006, 168) to think that Armstrong's theory must be inadequate because he cannot explain the contingency of the relations between particulars and properties (because the particulars are defined by their properties). However, Armstrong (2004, 81) does perhaps hint at an answer to this worry when he says that although the relations between particulars and properties are in some sense necessary (they couldn't occur without being so related) it is nonetheless the case that

these particulars and properties are contingent. It could be that different but similar ones obtained instead.

13. The name 'slingshot argument' originates with Barwise and Perry (1981), but the argument appears to be originally attributed to Frege (1892/1980). Neal (1995) doubts the validity of this attribution. If the attribution is fair then the form of the argument given by Frege differs widely from the form of the argument as I will give it here. Neal (1995) considers a variety of forms. I am tacking my presentation largely from Searle (1995, appendix to ch. 9), Mellor (1999, 114), and Lowe (1998, 229–230, ftnt. 2).
14. Cf. Russell (1905, 1919, ch. XVI, 1968, lect. VI). I will say more about this in the next chapter.
15. Cf. Neal (1995) and Lowe (2002, 172–173).
16. Cf. Searle (1995, appendix to ch. 9) and Barwise and Perry (1981). Olson (1987, ch. 4) considers variants of both.
17. Armstrong thinks that this empirical criterion is insufficient on its own, as there might be necessarily concurrent states of affairs. Armstrong thinks that the constitutive identity criterion is sufficient on its own, however, Lowe (1998, ch. 11.5) raises doubts about this. I think that by combining these two criteria we go some way to alleviating these doubts. Further, as I am not wedded to Armstrong's ontology, I can borrow Lowe's own criteria for individuating states of affairs (by which he means something ontologically non-fundamental).
18. This example perhaps reflects more of an epistemological issue than a matter of practical rationality, however, considering it helps to clarify the issues.
19. Cf. Armstrong (1997, ch. 13) and also Armstrong (2004, ch. 6).
20. So when I say the states of affairs that *none* of them are apples this should not be taken as a negative, rather we need only think of the various positive states of affairs composed of green objects and properties of *being a ball*, *being a cabbage* etc. (The same can be said for the other cases below.) (This is not reliant on the fact that being a cabbage excludes being an apple or the fact that nothing can be both an apple and a cabbage. However, this relates to a proposal that has been put forward that instead of some negative facts, we could make do with positive facts which exclude these negative ones (cf. Demos (1917)). Armstrong (2004, ch. 5.2.1) considers such a view, but finds it insufficient because it depends too much on the contingent nature of reality for whether or not all nega-

tives are actually excluded (cf. Molnar (2000) and Rodriguez-Pereyra (2006)).)

21. Cf. respectively, Russell (1968, lect. III and V) and Martin (1996). Cf. Beall (2000) (and similarly Barwise and Perry (1983)) who offers a rather different proposal which offers negative facts without real negativity by instead positing simply positive facts of different polarities (though perhaps for this reason his proposal is dubious).

References

Armstrong, D. M. (1997). *A World of States of Affairs*. Cambridge: Cambridge University Press.

Armstrong, D. M. (2004). *Truth and Truthmakers*. Cambridge: Cambridge University Press.

Barwise, J., & Perry, J. (1981). Semantic Innocence and Uncompromising Situations. *Midwest Studies in the Philosophy of Language, 6*, 387–403.

Barwise, J., & Perry, J. (1983). *Situations and Attitudes*. Cambridge, MA: Massachusetts Institute of Technology Press.

Beall, J. C. (2000). On Truthmakers for Negative Truths. *Australasian Journal of Philosophy, 78*, 264–268.

Bradley, F. H. (1897). *Appearance and Reality: A Metaphysical Essay*. London: Swan Sonnenschein.

Braun, D. (2006). Illogical but Rational. *Nous, 40*, 376–379.

Campbell, K. (1990). *Abstract Particulars*. Oxford: Blackwell.

Davidson, D. (1973/2001). Radical Interpretation. In *Inquiries into Truth and Interpretation* (pp. 125–139). Oxford: Oxford University Press.

Demos, R. (1917). A Discussion of a Certain Type of Negative Proposition. *Mind, 26*, 188–196.

Frege, G. (1892/1980). On Sense and Meaning (M. Black, Trans.). In P. Geach & M. Black (Eds.), *Translations from the Philosophical Writings of Gottlob Frege* (pp. 56–78). Oxford: Blackwell.

Kaplan, D. (1989). Demonstratives. In J. Almog, J. Perry, & H. Wettstein (Eds.), *Themes from Kaplan* (pp. 481–563). Oxford: Oxford University Press.

Kim, J. (1988). What is 'Naturalized Epistemology?' *Philosophical Perspectives, 2*(Epistemology), 381–405.

Le Poidevin, R. (2007). *The Images of Time*. Oxford: Oxford University Press.

Lowe, E. J. (1998). *The Possibility of Metaphysics*. Oxford: Oxford University Press.

Lowe, E. J. (2002). *A Survey of Metaphysics*. Oxford: Oxford University Press.
Lowe, E. J. (2006). *The Four-Category Ontology: A Metaphysical Foundation for Natural Science*. Oxford: Oxford University Press.
Martin, C. B. (1996). How It Is: Entities, Absences, and Voids. *Australasian Journal of Philosophy, 74*, 57–65.
Mellor, D. H. (1999). *The Facts of Causation*. London: Routledge.
Mollnar, G. (2000). Truthmakers for Negative Truths. *Australasian Journal of Philosophy, 78*, 72–86.
Neal, S. (1995). The Philosophical Significance of Godel's Slingshot. *Mind, 104*, 761–825.
Olson, K. R. (1987). *An Essay on Facts*. Stanford: Centre for the Study of Language and Information.
Pearson, O. (2015). Rationality and the First Person. *Journal of Consciousness Studies, 22*, 132–148.
Rodriguez-Pereyra, G. (2002). *Resemblance Nominalism: A Solution to the Problem of Universals*. Oxford: Oxford University Press.
Rodriguez-Pereyra, G. (2006). Truthmakers. *Philosophy Compass, 1*, 186–200.
Russell, B. (1905). On Denoting. *Mind, 14*, 479–493.
Russell, B. (1919). *Introduction to Mathematical Philosophy*. London: Allen & Unwin.
Russell, B. (1968). The Philosophy of Logical Atomism. In R. C. Marsh (Ed.), *Logic and Knowledge: Essays 1901–1950* (pp. 175–282). London: Allen & Unwin.
Salmon, N. V. (2006). The Resilience of Illogical Belief. *Nous, 40*, 369–375.
Schiffer, S. R. (2003). *The Things We Mean*. Oxford: Oxford University Press.
Schiffer, S. R. (2006). A Problem for a Direct-Reference Theory of Belief Reports. *Nous, 40*, 361–368.
Searle, J. R. (1995). *The Construction of Social Reality*. London: Penguin.
Simons, P. (1992). Logical Atomism and Its Ontological Refinement. *Philosophical Studies, 51*, 157–179.
Skyrms, B. (1981). Tractarian Nominalism. *Philosophical Studies, 40*, 199–206.
Soames, S. (2002). *Beyond Rigidity: The Unfinished Semantic Agenda of Naming and Necessity*. Oxford: Oxford University Press.
Wason, P. C. (1966). Reasoning. In B. M. Foss (Ed.), *New Horizons in Psychology* (Vol. I, pp. 135–152). Harmandsworth: Penguin.
Wittgenstein, L. (1921/2001a). *Tractatus Logico-Philosophicus*. London: Routledge.
Wittgenstein, L. (1953/2001b). *Philosophical Investigations* (G. E. M. Anscombe, Trans.). Oxford: Blackwell.

8

Implications for Non-indexical Cases

In the last chapter I argued that the need for tensed and first-personal beliefs in rational actions entails that these beliefs capture facts not captured by other beliefs—the argument from action. In this chapter I will show how my argument differs from issues of cognitive significance, such as Frege's Puzzle. Frege's Puzzle can show that a tensed and a tenseless belief must differ in cognitive content and hence that they must be different beliefs. But my argument goes further by showing that this difference in beliefs relates to a difference in the actions the beliefs make rational and hence to a difference in the facts that they concern.

This raises a question that I will discuss in this chapter: might a variant of the argument from action apply to different beliefs? If rationality demands one belief in place of another, then the two beliefs must involve an awareness of different reasons and hence facts. I will show that my argument from action is applicable to other cases, but that this is unproblematic because it simply fits with a common practice of explaining a difference in cognitive significance in terms of a difference in descriptive content. Further, unlike the case with tensed and first-personal beliefs, the new cases tells us little about our ontology or metaphysics because any required facts are of a familiar form.

Olley (F.O.C.H.) Pearson, *Rationality, Time, and Self*,
https://doi.org/10.1007/978-3-319-71973-3_8

In the course of this discussion it will appear that the argument from rationality presupposes a fine grained notion of actions. I will show that it does not.

8.1 Cognitive Significance

Identity statements are often trivial and uninformative. For example, it is generally trivial and uninformative to be told that "Dodgson is Dodgson". However, sometimes identity statements are informative. It is often informative to be told that "Dodgson is Carroll". But, if both of these identity statements are true, one might wonder why the latter is informative. This is the core of the puzzle concerning the cognitive significance of proper names that I will refer to as *Frege's puzzle.*[1] If one identified the meaning of a name simply with its referent, then the names 'Dodgson' and 'Carroll' would have the same meaning. If one identified the cognitive significance of an utterance with its meaning, then it would follow that the two identity statements would have the same cognitive significance. However, the example shows that they do not. The implication of the puzzle is that either the meaning of a name is not simply its referent or the cognitive significance of a name is not simply its meaning. (I will call the idea that the meaning of a name is its referent the *Millian* view, because Mill offered a view like it in regard to names.[2])

Very similar puzzles can also be shown to arise for expressions other than names. For example, the two definite descriptions *the author of Euclid and his Rivals* and *the author of Alice's Adventures in Wonderland* denote the same person, namely Carroll and hence Dodgson. But, it is generally trivial to be told that "the author of Euclid and his Rivals is the author of Euclid and his Rivals" but it is informative to be told that "the author of Euclid and his Rivals is the author of Alice's Adventures in Wonderland". If one took the meaning of a definite description to be its denotation and the meaning of an expression to give its cognitive significance, then one would expect these two utterances to share cognitive significance. But, because one generally appears trivial whilst the other does not they do not share cognitive significance.[3]

These puzzles immediately link with a second puzzle. If co-referring or co-denoting expressions can have different cognitive significance, then utterances describing beliefs ought to pay heed to this fact. Inter-substitution of co-referring/co-denoting expressions often fails in such utterances. For example, if Jo does not know that Carroll and Dodgson are the same person, then Jo might believe that Carroll is making a mess but not that Dodgson is. Jo believes Dodgson to be dark haired and careful man from when she was introduced to him at work. She believes Carroll to be blond and scruffy from when she met him at a party where he was in fancy-dress. So whilst it is true to say "Jo believes that Carroll is making a mess" it is false that "Jo believes that Dodgson is making a mess". Explaining why substitution fails in these cases is the *puzzle of substitution*. It was raised by Frege (1892/1980) and exercised by Russell and many others since.[4] I will refer to both puzzles together as the puzzles of *cognitive significance*. (I said that substitution usually fails in utterances describing beliefs because there is a way of understanding such utterances according to which failure does not occur. There are so called *de re* beliefs. For example, we might say that Jo believes of the person that is Dodgson that he is making a mess. This is taken by most to be a different case and I will not focus on it here. Though it is of note that Perry's argument shows that *de re* beliefs do not do away with the essential indexicals.[5])

The problem of the essential indexical as raised by Perry could be seen as a further puzzle of cognitive significance. For example, if I am Francis and I can fail to know that I am, then it appears that the two expressions 'I' and 'Francis' will have different cognitive significance for me. Moreover, if I must believe "I am making a mess" rather than "Francis is making a mess", then perhaps this is because the two beliefs have different cognitive significance for me. Similarly, even if it is now 2pm, I might not know that and so 'now' and '2pm' will have different cognitive significance for me.

If one was an ardent Millian, then one might take the puzzles of cognitive significance to show that many apparently co-referring names in fact do not co-refer. Carroll is not Dodgson and this is why "Dodgson is Dodgson" is trivial but "Dodgson is Carroll" is not trivial (but in fact false). Similarly one might insist that many apparently co-denoting definite descriptions do not co-denote. This would be a rather implausible conclusion to draw. The examples I have already given imply that a person can have two names and yet another can fail to know this. Similarly, a person can fulfil more than one definite description.

In short, the puzzles of cognitive significance appear not to warrant the conclusion that a person cannot have two names. Noting the link between the puzzles of cognitive significance and the case of the essential indexical might hence lead one to say that we ought not to be led by this latter case to conclude that my belief 'I am making a mess' concerns different facts from my belief 'Francis is making a mess'. Similarly we should not to be led to conclude that my 2pm belief 'I must leave now' concerns different facts from my tenseless belief 'it must be the case that I leave at 2pm'. However, it is important to notice that the argument from action goes far beyond an issue of cognitive significance. The point of my discussion of reasons and rationality is to show that it is not enough to explain the essential role of indexical beliefs to say that they have a different cognitive significance to non-indexical ones. Rather, a rational demand for an indexical belief must be a demand for a belief which is an awareness of a particular fact. Undoubtedly such a belief will also have a different cognitive significance. But, if indexical beliefs and non-indexical beliefs merely differed in their cognitive significance, then rationality would not require one in place of the other. Therefore, we should not equate the argument from action with the puzzles of cognitive significance.

8.2 The Application of the Argument from Action to Non-indexical Cases

With the puzzles of cognitive significance in mind one might wonder if the argument from action can be applied to other cases. For example, if two co-referring names can have different cognitive significance, then might an agent be rational to act on a belief involving one name and not on a belief involving a co-referring one?

Before considering this question I should say something of the assumptions of my discussion. In addressing the above question I will spend some time looking at theories of meaning or semantics. These theories clearly tell one primarily about language and uses of language, about sentences and utterances, whilst my argument concerns beliefs. However, I have characterized beliefs partly in terms of a propensity to accept or assent to certain utterances or sentences. It follows that if utterances/sentences differ in meaning, then the associated beliefs will also differ. Further, if grasping the meaning of an utterance/sentence involves grasping certain facts, then those facts must also be grasped in the corresponding beliefs. Therefore, a discussion about semantics will aid a discussion of beliefs. Moreover, it appears plausible that utterances/sentences with different meanings will have different cognitive significances even if we do not simply identify cognitive significance and meaning. So, a focus on semantics is again legitimate.

Suppose that a person has two names: 'Harry' and 'Will'. Further, suppose that although I am familiar with both names I do not know that they denote the same person. If I am told that Harry is in danger, then I might form the belief 'Harry is in danger' and hence be aware of a reason to help Harry. In such a case it seems to be rational for me to form the goal giving desire to help Harry. But, it does not appear to be rational for me to form the goal giving desire to help Will because for all I know Will is perfectly safe. Alternatively, if I formed the belief 'Will is in danger', then it would appear rational for me to adopt the goal of helping Will but not necessarily of helping Harry. It thus seems as if rationality can demand a particular name involving belief in place of a different co-referential name involving belief. It follows that the argument from action ought to apply in this case too.

Similar considerations arise for the case of definite descriptions. It is possible that one person both fixed my car last summer and lives next door to me now. Further, it is possible that I do not know that one person satisfies both of these definite descriptions. If this is the case then it would seem that there is something irrational in my forming a goal of helping the person who lives next door to me now because I form the belief 'the person who fixed my car last summer is in danger'. Conversely, it seems irrational of me to adopt the goal of helping the person who fixed my car last summer because I form the belief 'the person who lives next door to me now is in danger'. Again, it appears that rationality might demand one definite description involving belief in place of another co-denoting definite description involving belief. Therefore, the argument from action also ought to apply in this case.[6]

However, things are not quite as simple as just suggested. In considering Perry's cases we start by considering the action and work back to the belief. Tom stops and in order for this action to be rational he must have a first-personal belief. If we follow the same pattern here it is less clear a belief involving one name rather than another will be required. For example, suppose the action in question is my running next door to help the person there in danger. Plausibly I must have a belief that the person next door is in danger, but my belief needn't refer to that person in any particular way. For example, it is equally rational if I believe 'Harry is in danger', 'Will is in danger', 'my neighbour is in danger', or 'my mechanic is in danger'. This sort of case is thus quite unlike the cases of the essential indexical. Tom must have a belief that refers to himself in a first-personal manner (and Mike must have a belief that picks out the time with tense).

Nevertheless, there does remain an issue here. Any belief that refers to the individual next door will do. But, it is not the case that any combination of beliefs will do. I cannot believe 'Harry is in danger' and 'Will is next door' if my action is rational. I would have to believe 'Harry is in danger' and 'Harry is next door' or 'Will is in danger' and 'Will is next door', for example. Therefore, there is a lingering sense in which the argument from action might apply to beliefs involving different co-referring names or co-denoting definite descriptions.

This places a certain demand on the argument from action. Rational actions are actions one chooses to perform for a reason for that action. Belief's sole role in this is to provide an awareness of the reason at issue. Similarly, the goal an agent has in a rational action is simply to perform that action. The belief must be an awareness of a fact which is a reason to perform the act which one has as a goal. It follows that my belief 'Harry is in danger' must be an awareness of a different fact from my belief 'Will is in danger', because the former, unlike the latter, can combine with my belief 'Harry is next door' in my rationally heading next door to help. Conversely, it appears—though I will subsequently deny this—that the actions I adopt as goals in desiring to help Harry and desiring to help Will must also be different because the former, unlike the latter, can combine with my belief 'Harry is in danger' in a rational action. More generally, if two beliefs involving different co-referring names can play different roles in rational actions, then they must concern different facts. It also appears that if two desires involving different co-referring names can play different roles in rational actions, then they must concern different actions. Further, if two beliefs involving different co-denoting definite descriptions can play different roles in rational actions, then they must concern different facts and if two desires involving two different co-denoting definite descriptions can play different roles in rational actions, then apparently they must concern different actions. In the following sections of this chapter I will argue that this is unproblematic and in fact independently plausible. I will first address the case with beliefs and facts, and subsequently that with desires and actions.

8.3 Co-reference & Distinct Facts: Names & Definite Descriptions

Frege's (1892/1980) answer to the puzzles of cognitive significance was to turn to the mode of presentation of the referent. This is the *sense* of the expression. Each referring expression was to have a sense as well as a referent. The sense of a sentence containing such referring expressions would be composed by the senses of these expressions and would give the *thought* captured by this sentence. This thought may be considered the proposition expressed by the sentence. The complete sentence itself will refer to either *The True* or *The False*, unless it is embedded within a further sentence in which case it will refer to its usual (i.e. unembedded) sense.

Suppose that the sense expressed by the name 'Dodgson' was *the author of Euclid and his Rivals* and that the sense expressed by 'Carroll' was *the author of Alice's Adventures in Wonderland.* These were the guises under which the names presented their referents. It would follow that the thought expressed by an utterance of "Dodgson studied at oxford" would be the thought 'the author of Euclid and his Rivals studied at oxford'. It would therefore be quite distinct from the thought expressed by "Carroll studied at Oxford" which would be the thought 'the author of Alice's Adventures in Wonderland studied at Oxford'. Nevertheless, the two sentences would designate the same referent, namely The True. The names composing those two sentences would also designate the same referent, namely the man who wrote both Euclid and his Rivals and Alice's Adventures in Wonderland. Thus Frege's puzzle of identity is answered because the utterances "Carroll is Carroll" and "Carroll is Dodgson" share truth-value and referent. But, they nevertheless differ in meaning because they have different senses and therefore they will differ in cognitive significance.

In describing beliefs we embed one sentence within another. For example, the sentence 'Carroll is making a mess' stands alone as a complete expression with a truth-value. However, it is emended in another sentence when we say "Jo believes that Carroll is making a mess". When a sentence is thus embedded Frege says that it refers to the sense it would have if it was not embedded. Because of this the two sentences 'Carroll is making a mess' and 'Dodgson is making a mess' will share truth-value and referent when they are not embedded. But, when they are embedded their referents become their usual sense. Unembedded these two sentences have different senses. Therefore, once embedded they have different referents. Thus it is clear that we shouldn't expect to be able to substitute the name 'Carroll' occurring within one for the name 'Dodgson' occurring within the other. The puzzle of substitution is avoided.

Frege's solution applies in a similar way to the case of definite descriptions. The definite description *the author of Alice's Adventures in Wonderland* will have the same denotation as the definite description *the author of Euclid and his Rivals*. But, the two will have different senses. In fact, in this case it is plausible to assume that the former definite description will share a sense with the name 'Carroll' if it captures the sense of that name. The definite description and the name alike will present Carroll as the author of Alice's Adventures in Wonderland.

Russell (1905, 1919, ch. XVI, 1968, lect. VI) offers a different solution to the puzzles of cognitive significance based on an analysis of definite descriptions such as *the author of Alice's Adventures in Wonderland*. He takes an utterance using a definite description to make a number of claims. For example, "The author of Alice's Adventures in Wonderland is making a mess" means that at least one thing authored Alice's Adventures in Wonderland, at most one thing authored Alice's Adventures in

Wonderland, and whatever authored Alice's Adventures in Wonderland is making a mess. According to Russell definite descriptions have the logical form—even if not a surface grammatical form—of quantified sentences in which the definite description does not appear.[7] Thus, definite descriptions do not refer although they do denote whatever object uniquely fulfils the description. This enables us to see that co-denoting definite descriptions can mean quite different things. For example, "The author of Euclid and his Rivals is making a mess" will mean there is at least one thing that wrote Euclid and his Rivals, there is at most one thing that wrote Euclid and his Rivals, and whatever wrote Euclid and his Rivals is making a mess. So the definite descriptions have quite different meanings and hence cognitive significance even though they co-denote.

Russell argues that all apparently referring expressions (except for the indexicals 'this' or 'that' and 'I') are in fact disguised or abbreviated definite descriptions.[8] Because of this Russell's analysis of definite descriptions can also deal with the problems of cognitive significance as they arise concerning names. For example, the name 'Carroll' will in fact express a definite description which denotes Carroll, perhaps *the author of Alice's Adventures in Wonderland.*[9] The name 'Dodgson' will express a different definite description though one that also denotes Carroll, perhaps *the author of Euclid and his Rivals.* Therefore, the two names will mean quite different things despite sharing a denotation.

I will refer to Frege's and Russell's theories as *descriptive theories.* They are theories of the way words refer or denote. Both of these theories encourage the view that sentences containing different co-referring/denoting names can be expected to capture different facts. This is because the different senses or definite descriptions associated with different names capture different facts. For example, the fact that someone wrote Alice's Adventures in Wonderland is quite distinct from the fact that someone wrote Euclid and his Rivals. Using Russell's analysis of definite descriptions[10] it is clear that "Carroll is in Oxford" and "Dodgson is in Oxford" capture different facts. The former will (we are supposing) capture the fact that someone both wrote Alice's Adventures in Wonderland and is in Oxford. The latter will capture the fact that someone both wrote Euclid and his Rivals and is in Oxford. To this degree the argument from action is vindicated. (Notice that Frege's and Russell's theories have independent support, so utilising them to support the argument from action is not an *ad hoc* manoeuvre.)

However, both Frege's and Russell's theories have come under attack, most noticeably from Kripke.[11] Kripke (1981) criticized Frege's and Russell's views by arguing as follows: K1, a person needn't be able to associate a definite description with a name to be able to refer with it; K2, being able to associate a definite description with a name is not sufficient to enable someone to refer with it; and finally, K3, definite descriptions and names have different logical shapes, that is they act differently in modal contexts.

Kripke gives argument K1 by considering a case in which a person appears to be able to use a name meaningfully despite not knowing anything uniquely individuating of the referent of the name. The example Kripke gives concerns Richard Feynman. Many people can use his name and know something about who they are referring to. For example, they may know that he was an American physicist. However, many people know little more than this about him. Crucially, they do not appear to know enough about him to distinguish him from Gell-Mann who was another famous American physicist. Yet despite this, it is clear these people can talk meaningfully about Feynman by using his name. It appears to follow that the name is not associated with a definite description for these individuals. Therefore, this cannot be the mechanism through which the name refers/denotes.

Kripke gives argument K2 by considering a case in which a person associates a definite description with a name, but the description applies to someone other than the referent of the name. Godel is commonly taken to have been the first person to have proven Incompleteness (a definite description). Kripke invites us to imagine that Godel in fact copied the proof from little known Schmidt. In such a case it would seem that we all still refer to Godel in using his name even if Schmidt is the person who is captured by the definite description associated with the name. More generally it seems that we can be mistaken about the things we refer to with names. It thus appears that associating a definite description with a name is not sufficient for referring/denoting with that name, contra Frege and Russell.

Kripke gives argument K3 by distinguishing between rigid designators and non-rigid designators. The former, unlike the latter, pick out the same individual in all possible worlds. For example, although Dodgson wrote Alice's Adventures in Wonderland it is true to say "It is possible that Dodgson was not the author of Alice's Adventures in Wonderland". Whereas, it is false to say "It is possible that Dodgson was not Dodgson". This is because the name Dodgson picks out the same person in every possible world. However, a definite description, such as *the author of Alice's Adventures in Wonderland*, will pick out different individuals in different possible worlds. For example, in a world where Dodgson did not write that book, but instead Orwell did, then the definite description will pick out Orwell not Dodgson. In short, names appear to be rigid designators whilst definite descriptions appear to be non-rigid designators. So, it makes little sense to suppose the latter can give the meaning of the former.[12] A second aspect of this problem for the descriptive theories is that if they were correct, then an utterance such as "if Dodgson exists then they wrote Euclid and his Rivals" would be necessarily true and a priori. Because, it would mean something like *if the author of Euclid and his Rivals exists then they wrote Euclid and his Rivals* which is necessarily true and knowable a priori. (It would mean this if, as we are supposing, that definite description gives the sense of the name.) But, the utterance "if Dodgson exists then they wrote Euclid and his Rivals" appears to be neither necessary nor knowable a priori. I'll call this *K3i*.

Despite their strong appearance Kripke's criticisms against the descriptive theories have all been responded to. These responses have largely centred on the idea that Kripke focused (as perhaps Russell and Frege encouraged) on the wrong descriptions.

It has been argued that there will always be a definite description of the referent of a name available to a speaker, contra to K1. For example, users of the name 'Feynmann', can associate with this name the definite description *the person named 'Feynmann'*, or the description *the person who is the causal origin of the information that I associate with the name 'Feynmann'*.[13] Such definite descriptions will concern Feynmann and will always be available to users of his name.[14] Jackson[15] has in fact argued for the necessary availability of descriptions from data used by Kripke himself. In K3 Kripke makes use of our ability to judge what a name refers to in different possible worlds. Jackson argues that this ability itself reveals that users of a name have in mind a stock of properties that they associate with bearer of the name and that they utilize to make these decisions.[16] A disjunction of such properties could be used to form a definite description capturing the referent of a name.[17]

K2 is in part responded to by the response to K1 because there it is shown that a speaker will always be able to associate an accurate definite description with a name that he or she uses. All that is required in addition is to distinguish this accurate description from other descriptions a speaker might associate with a name. If the description at issue is of the form indicated by Jackson, then this distinction is accomplished. Because, only the (disjunctive) definite description that a speaker takes to pick out a referent when considering different possible worlds need be taken as that which plays a crucial role in determining the referent/denotation of a name.

Kripke himself notes that some definite descriptions are rigid designators. For example, the definite description *the positive square root of four* will denote two in every possible world. More generally, other definite descriptions can be rigidified by including a modifier such as 'actually'. For example, *the actual author of Euclid and his Rivals* will pick out Dodgson in every possible world. Because, it remains the case that in the actual world Dodgson was the author whatever world we consider.[18] It follows that insisting that names are rigid designators needn't rule out definite descriptions from providing the meaning of names. All that it does appear to rule out is non-rigid definite descriptions providing the meaning of a rigid name. Dummett (1973, 110–151) instead emphasizes a point noted by Russell[19]: a definite description can be interpreted to take wide scope in modal contexts. For example, we can interpret "It is possible that Carroll didn't visit Oxford" in two ways. If we take the description to have narrow scope the utterance means something like *it is possible that exactly one thing wrote Euclid and his rivals and it did not visit Oxford*. In this interpretation the utterance will be true if and only if there is a possible world in which someone authored Euclid and his Rivals and did not visit Oxford. Whether or not Carroll was the author in that world would make no difference. In essence, the definite description will act akin to a non-rigid designator. However, if we interpret the utterance in such a way that the definite description has wide scope we end up with something like the following: *there is exactly one thing which wrote Euclid and his Rivals and it is possible that that thing did not visit Oxford.* This will only be the case in worlds in which Carroll did not visit Oxford. Whether or not he wrote Euclid and his Rivals in that world would make no difference. Hence this wide scope interpretation will give us the results Kripke's argument required.[20]

Closing the gap between definite descriptions and rigid designators in this way undermines several aspects of K3. However, more must be said to deal with K3i or the appearance that "if Carroll exists then they wrote Alice's Adventures in Wonderland" will be necessarily true and knowable a priori.

The lesson from the response to K1 is that the descriptions that give the meaning of names will not be of a form like *the author of Euclid and his rivals.* In fact, bearing in mind what has been said about rigidity we might take the description associated with the name 'Carroll' to be more like *the person actually referred to by the name 'Carroll'.* The supposed problematic sentences raised in K3i would thus be of the form 'if Carroll exists then they are the person actually referred to by the name "Carroll"'. However, this sentence is plausibly both necessary and a priori as the descriptivist[21] would want in the face of K3i. The rigidifying nature of 'actually' means that if Carroll has actually been referred to by the name 'Carroll' (which he has), then it is necessary that he has. There is no counterfactual situation or possible world the consideration of which will show that in the actual world Carroll has not been referred to by that name. The idea that this can be known a priori might be less obvious. For example, one might wonder how Carroll's mother can have known he would be called 'Carroll' the first time she used his name. But, if this shows anything it is simply that the description being considered cannot have been the meaning of the name when it was first used. This could be because the name simply lacked a meaning until it was first used and introduced into the language. Alternatively, perhaps we are simply dealing with the wrong description.[22] If Jackson is right, then there will be another description available. Further, if this other description plays the role of defining the modal nature of the name (picking out its referent in different possible worlds), then it must be that the object fits this description of necessity.[23]

In this way I take it that descriptive theories or variants thereof can be made plausible despite attacks such as Kripke's. If a descriptive theory of some sort is correct, then this means different names will have different information associated with them. They will capture different facts. For example, the fact that Carroll is actually called 'Carroll' is a different fact from the fact that he is actually called 'Dodgson'. To this extent the argument from action is vindicated even if rationality can demand one name involving belief in place of another different co-referring/co-denoting name involving belief.

Nevertheless, a defence of the argument from action does not require the adoption of a descriptive theory of names. Because, such a theory is not the only theory that supports the idea that sentences differing only over their containing different co-referring/denoting names will capture different facts. Rather, many theories make use of this to account for the puzzles of cognitive significance. Besides the idea that the descriptive/informational content is part of the meaning of a sentence in the sense intended by descriptive theories one might instead say it is part of the semantic content but plays a different role[24] or one might say it is part of the pragmatic content.

For example, in defending a non-descriptive theory Soames[25] suggests that co-referring names[26] will have the same semantic content, namely their referent. But, one name will often be used rather than another co-referring one because they will express different information or different descriptive content. In saying this Soames takes himself to be furthering and also correcting the thoughts of Kripke. However, I will not go into this non-descriptivist theory in detail here.

One important difference between Soames' theory and the descriptive theories discussed is that Soames takes the difference in information captured by different co-referring names to be in some sense context sensitive.[27] This might lead one to worry whether it can fulfil the role I need. However, this difference in descriptive content is introduced in part to deal with the puzzles of cognitive significance and it is precisely when one takes two names to have a different cognitive significance that one might take the argument from action to apply to the case of name involving beliefs. It follows that even if Soames does not think that a difference in descriptive content will always arise for co-referring names, he will take it to arise in all cases where the argument from action takes it to.

More generally, one might be tempted to think that co-referring names share a semantic value but that the puzzles of cognitive significance can be explained simply by individuals associating different information with different co-referring names because they are ignorant of the co-reference.[28] If one adopted this approach to cognitive significance, then this simple theory ought also to accommodate the argument from action. Because, this simple theory already implies that the agent will be capturing different descriptive information—facts—when having beliefs involving different co-referring names. (I admit the exact way this further content is involved is here left unclear. But, it is clear that it must be involved in a strong way as it effects cognitive significance: the further facts impinge upon the persons conscious states of awareness.)

Thus, it is plausible to think that beliefs involving different co-denoting definite descriptions will capture different facts. Because, the beliefs will, in accord with Russell's theory of descriptions,[29] pick out different facts as being the case.[30] In part because of this, it is also plausible to think that beliefs involving different co-referring or co-denoting names will capture different facts in a way that vindicates the argument from action. This follows straightforwardly if one adopts a descriptive theory of names and this is a view with independent plausibility that continues to hold supporters. However, it also follows if one does not adopt a descriptive theory but instead follows Soames or the general temptation to explain cognitive significance in terms of information associated with an expression.

8.4 Natural Kind Terms & the Argument from Action

It is possible that the argument from action applies to beliefs involving co-referring/denoting names and definite descriptions and in a similar way it is also plausible that it can apply to beliefs involving co-denoting/

referring natural kind terms. For example, it could be that the terms 'water' and 'H_2O' both denote the natural kind water. However, an individual might not know this and so the two terms may have different cognitive significance for such an individual. It would appear to be rational for a person like this to adopt the goal of drinking water after forming the belief that drinking water is good for his or her health. But, it would be irrational of that person to adopt the goal of drinking H_2O after forming that belief because for all he or she knows this is a highly poisonous substance used by chemists.

It follows that the argument from action could also apply to beliefs involving different co-denoting kind terms and hence that in order to make the argument from action plausible it must be plausible that such beliefs can capture different facts. I will not defend this idea in great detail. But, the discussion of kind terms overlaps the discussion of names. Many of the same issues arise alike in both fields and many writers cover the two cases side by side.[31]

The puzzles of cognitive significance can be recast using kind terms. For example, it is often trivial to be told "water is water" but informative to be told "water is H_2O". One can suggest kind terms have senses as well as referents. Alternatively, one can suggest that kind terms do not refer but that they denote by being disguised definite descriptions. If these ideas are correct, they can account for the puzzles of cognitive significance. Nonetheless, to make these ideas plausible they must be detailed in such a way as to account for Kripke's criticisms. But, Kripke's worries can be met in the case of kind terms. For example, water might be associated with the rigidified definite description *the stuff actually referred to as 'water'*.

Nevertheless, introducing these issues to natural kind terms too raises a further problem for the descriptivist approach. According to descriptivism an expression refers/denotes by being associated with a definite description. But this of course raises the question of how the terms in the definite description themselves refer/denote. It could be that they also refer/denote via associated definite descriptions. However, we cannot carry on with this pattern (e.g. for the reference of the terms of these further descriptions, and so on), otherwise we would be launched on an infinite regress or around in a circle.[32] This would appear to entail that the referent/denotation of a term could never be established.[33]

One way to answer this worry is to suggest that there are a number of terms whose reference/denotation are not fixed by definite descriptions. Chalmers (2004, 2006) defends an idea like this by speaking of *semantically neutral* or *qualitative* terms. These are terms for which the two-dimensions (reference to possible worlds as counterfactual and as actual) would provide no difference in reference/denotation. Chalmers (2004, 191) gives the following terms as examples: 'and', 'philosopher', 'friend', 'consciousness', and 'cause'.

An alternative response is provided by Jackson (1998b) who argues that descriptivism needn't suppose that the associated descriptions are linguistic. Jackson speaks of the reference/denotation of a term being fixed by the language user associating properties with the term rather than saying that it is fixed by a description. The referent/denotation is the object that possesses these qualities. This makes it clear that the way that these qualities are thought of can be quite distinct from the way that the term refers/denotes. For Jackson the descriptive theory is a theory of language. It is a theory of the meaning of words like 'water'. There is no need to suppose that words which have their meaning given to them in part by practice will refer in the same way as the components of thought which appear not to have their meaning given to them by practice. Thus there is no infinite regress.

It is plausible to think that a difference in associated properties—in Jackson's sense—is sufficient to give a difference in captured facts as is my concern. Because, the fact that an object has one property is different from the fact that it has another one. Further, although Jackson draws a distinction between thought and language, and my primary concern is with thought, this does not undermine my use of the descriptive theory. My concern is that, for example, the thoughts 'Carroll is in Oxford' and 'Dodgson is in Oxford' capture different facts. I am concerned to defend this idea by arguing that the utterances "Carroll is in Oxford" and "Dodgson is in Oxford" capture different facts, and by arguing that this shows that the two thoughts must capture different facts because the thoughts are recognized in part with a propensity to accept or make such utterances. This is in no way reliant on supposing that the thoughts are linguistic.

I do not mean to assert that natural kind terms and names ought to be treated as the same in all respects. But, I hope to have made it clear that it is plausible to think that beliefs involving different co-denoting natural kind terms will capture different facts just as those concerning different co-denoting names will.[34] Therefore, it is not problematic if the argument from action is applicable to them.

8.5 Actions

Above we saw that an agent might act rationally whilst believing 'Harry is in danger' and having the goal of helping Harry but not in believing 'Will is in danger' and having the goal of helping Harry even if Will is Harry. According to the argument from action it must follow from this that the two beliefs pick out different facts and I have just defended this idea. Similarly, an agent can act rationally in believing 'my neighbour is in danger' and forming the goal of helping his or her neighbour but not act rationally in believing 'my mechanic is in danger' and forming the goal of helping his or her neighbour even if the person's neighbour is his or her mechanic. Again, according to the argument from action it must follow from this that the two beliefs pick out different facts and I have also defended this idea. The reason that the argument from action had these implications is that rationality will only demand one belief in place of another in a rational action if the two beliefs pick out different reasons—facts.

We also saw that an agent can act rationally when he or she believes 'Harry is in danger' and forms the goal of helping Harry but not when he or she believes 'Harry is in danger' and forms the goal of helping Will. Similarly, when he or she believes 'my mechanic is in danger' and forms the goal of helping his or her mechanic an agent can be rational, but not when that agent believes 'my mechanic is in danger' and forms the goal of helping his or her neighbour. One might suppose that the sole requirement on a goal giving desire in a rational action is that the goal is to perform the action which is justified by the reason the agent is aware of through his or her beliefs. If this was the case, then it would seem to follow that because the former beliefs in the two cases just outlined concern the same reasons it must be that the two goals in each case concern different actions. It would appear that if this was not the case rationality would not differentiate between them in this way. In this section I will argue that this is mistaken. A rational agent must act for reasons that justify that action. This requires that the agent has a goal giving desire which is to perform that action. So, there is a rational requirement that the goal is to perform the action for which the agent has reasons. But, plausibly there are additional rational requirements on the goal. Therefore, an agent might be rationally required to have one goal in place of another even though both of those goals concern the same action. Before arguing this I will say something about the individuation of actions to show that it might not be problematic to say that the action of helping Will was different from the action of helping Harry.

The individuation of actions has divided authors. Some believe that there are only very coarse grained distinctions—*unifiers*.[35] Others believe that there are very fine grained distinctions—*multipliers*.[36] Adopting Anscombe's (1957/2000, sect. 23, p. 3) lead we might consider the following case: Bart is moving his arm up and down; he is thereby working the arm of a pump and pumping water to a house; the water has been poisoned; so, Bart is also thereby poisoning the people in the house and thereby killing those people. The question of action individuation I am here concerned with is the question of how many actions Bart is performing in this case.

One might argue that Bart is only performing one action, perhaps moving his arm.[37] The circumstances in which this action occurs are such that Bart thereby also pumps water because he is at a pump, Bart poisons because the pump is connected to a poisoned water supply which is being drunk, and Bart kills because the people drinking the water supply end up dead as a result. However, Bart doesn't need to do anything besides moving his arm in order to pump, poison and kill. It just happens that the circumstances are such that in moving his arm he is pumping, poisoning and killing, and therefore these are all one action of Bart's.[38]

However, one might argue that the arm moving, pumping, poisoning and killing are all different actions. This is because different things appear to be true of each. For example, it is true to say that Bart kills by poisoning, but false to say that Bart poisons by killing and odd to say that Bart poisons by poisoning. The 'by' relation appears to be asymmetric and irreflexive, but no asymmetric or irreflexive relation can hold between a thing and itself.[39] It also appears that these actions stand in different causal relations. For example, it seems that Bart's pumping caused water to flow, but it seems false to say that Bart's killing caused water to flow.[40]

A second type of consideration also supports the multipliers' idea that Bart is performing numerous actions: the actions appear to happen at different times. For example, killing involves someone dying. So, it appears that Bart cannot have killed anyone until that person dies. However, it might be that the people Bart kills do not die until after he stops moving his arm or pumping. Therefore, pumping and arm moving are not occurring when killing is and thus they must be distinct actions from killing.

Unifiers can try to respond to these arguments for taking Bart to be doing numerous actions. In the first case one might argue that in saying "Bart poisons by killing" one is not picking out a relation between two actions because 'killing' does not pick out a specific action because it is not a referring/denoting expression. (The expressions 'the killing Bart performs in 1957' or 'Intention-gate' would be referring/denoting.)[41] Nevertheless, there remains the implication that Bart fulfils one action type by fulfilling another. Therefore, if fulfilling an action type is embodying an action token and embodying an action token is acting, then we still have the implication that the actions are distinct.

One might respond to the multipliers' causal argument by distinguishing between a context in which one describes causal relations and a context in which one provides an explanation in terms of causes. We would expect the former context to be extensional so any expression referring to *c* could replace the '*c*' in "*c* caused *e*" without changing the truth-value of the sentence. However, we would not expect the explanatory context to be extensional. This would enable one to say that when considering "Bart's killing caused water to flow" we must bear the context in mind. If we take the context to be one in which we are simply picking out causal relations, then this sentence is strictly true. But, the sentence would provide a poor explanation and so in the explanatory context would not play the same role as "Bart's pumping caused water to flow". It is because "Bart's killing caused water to flow" would provide a poor explanation that it would be an odd thing to say. But, this does not mean that it is false. The distinction between contexts thus enables the unifier to insist that Bart only performs one action and to provide an error theory for the multipliers temptation to mistakenly say "Bart's killing caused water to flow" is false but "Bart's pumping caused water to flow" is true. Nevertheless, a multiplier might worry that even when one is aware that the context is extensional "Bart's killing caused water to flow" still sounds wrong.[42]

The unifier might respond to the argument from temporality by distinguishing between, for example, when an act becomes recognizably an act of killing and when that act occurs. If no one dies for a day after Bart stopped pumping, we would not say that Bart was killing for that day. Thus, it doesn't appear that we need to think of the act of killing as having this drawn out duration. Therefore, we are not prevented from identifying it with the act of pumping or of moving his arm. It is not the case that at the event of Bart's victim dying Bart stops killing. Rather, at the event of the victim dying it becomes clear that Bart's act of pumping was in fact an act of killing. At this later time we are able to describe the act as an act of killing. We could not do this at the earlier time. But, the act of killing happened when the pumping did and sometime before the death.[43]

However, the multiplier is also able to respond to the unifiers' argument that the agent only does one thing. Botting (2010) points out that we might say that someone is doing something when that person is not moving. For example, a person might be said to be printing despite being sat motionless at his or her computer waiting for the printer to finish. In such a case it appears that the agent's doing in selecting the print option on his or her computer is quite distinct from the agent's doing in printing because the former is over before the latter. This means that unifiers speak too quickly when they say that Bart needn't *do* anything further in order to poison other than he does to move his arm. Because, it is no longer clear that an agent must be in motion in order to be doing. Some cases differ from the printing case. For example, we are unwilling to say Bart is doing anything for the time in which the poison he has spread is taking effect on his victims. However, it needn't follow from this that Bart is not at this time acting or killing because the phenomena might have a different explanation. For example, it might be that we do not say that Bart is doing anything over this period because he has insufficient control over the spread of the poison. It is not the case that we do not describe him as doing anything because the killing which is one of his actions isn't occurring at this time.[44] In short, an unwillingness to describe an agent as doing or acting at a time does not entail that one of the agent's actions is not occurring at that time. It might simply follow from the nature of the action instead.

The point of the discussion of the preceding few paragraphs was to show that there are good reasons to adopt a fine grained approach to action individuation. This is a view with independent plausibility. If one did adopt the fine grained or multipliers approach, then it might plausibly follow that the action I had as goal in desiring to help my mechanic was distinct from the action I had as goal in desiring to help my neighbour. Further, it is possible that the action I had as goal when I desired to help Will was different from that which I had as goal when I desired to help Harry. This possibility is perhaps especially plausible if we adopt a descriptivist theory of names. It seems fine to say that I helped my mechanic by helping my neighbour or Harry by helping Will. But, it is questionable to say that I helped Will by helping Will or my mechanic by helping my mechanic. If this is right, then the argument from action would be vindicated.[45]

Nevertheless, as already suggested, I think there is an alternative way to vindicate the argument from action. I am only pressed to suppose that the goal of helping one's mechanic and the goal of helping one's neighbour concern different actions so long as we suppose that the sole requirement on a desire giving goal in a rational action is that it concern the reasonable action performed. But, this supposition can be denied. A desire giving goal must meet this requirement, but it must also meet a further requirement if the action is to be rational.

In Chap. 4 I pointed out that desires can be responsive to reasons. One can have a desire for a reason. I can want to go cycling because I gain pleasure from cycling. I now contend that in order to act rationally the agent's goal giving desire in the action must be appropriately responsive to the reason the agent performs the action for. Specifically, the reason the agent performs the action for must also be the reason for which the agent forms the goal giving desire that is the agent's state of motivation in acting. This does not mean that the agent must choose to have that desire. A desire's responsiveness to reasons is different from an action's responsiveness to reasons.

It follows that I can admit that in desiring to help my mechanic and in desiring to help my neighbour I will have the same action as goal and yet a rational action for which the first desire is the state of motivation must be responsive to different reasons to one for which the second desire is the state of motivation. An agent is rational to believe 'my mechanic is in danger' and to form the goal of helping his or her mechanic and act to help his or her mechanic for the reason he or she is aware of in this belief. The belief that 'my mechanic is in danger' captures a fact which is a reason to form the desire to help one's mechanic and a reason to act to help one's mechanic. However, an agent is not rational to believe 'my mechanic is in danger' and to form the goal of helping his or her neighbour and to act to help his or her neighbour for the reason he or she is aware of in this belief. This is because the belief that 'my mechanic is in danger' captures a fact which is not a reason to form the desire to help one's neighbour even if it is a reason to act to help one's neighbour.

It is quite plausible that the desire to help one's mechanic is different from the desire to help one's neighbour even if one's mechanic is one's neighbour. In fact, if this were denied, then the very worry I am trying to deal with in this section appears to evaporate. It is therefore quite possible there will be different reasons to adopt the two desires. Moreover, it is plausible that there are different reasons for the two just as is implied by the case at hand. If I do not know that my mechanic is my neighbour, then it is rational to be brought to form the goal of helping my mechanic by an awareness of the fact that my mechanic is in danger. But, this awareness does not make it rational to form the goal of helping my neighbour. This is because even if the desires concern the same actions, the desires are more fine grained than the actions and are receptive to the distinction between the fact that someone is a mechanic and in danger and the fact that someone is a neighbour and in danger. My preceding discussion of descriptive theories of names shows how this can also account for the difference in roles of the two desires to help Will and to help Harry. Different facts will be associated with the names 'Will' and 'Harry', and the two desires will be responsive to these different facts because they are as fine grained as the beliefs that involve those names are.

I can thus account for why the desire to help Harry might play a different role in the rationality of an action than the desire to help Will even if Will is Harry. Similarly, I can account for why a desire to help my mechanic might play a different role to a desire to help my neighbour even if my mechanic is my neighbour. The quick response to this is that the actions one adopts as goals in these two desires are different. This response relies on adopting a fine grained theory of action individuation and this is a view which has some independent plausibility. A further and distinct response notes that the goal giving desire, as one's state of being motivated in acting, must itself be appropriately responsive to reasons. An agent must desire for the reasons that agent acts for when that agent acts rationally. Desires admit of fine grained individuation (even if actions do not) and thus actions involving different goal giving desires may differ in their rationality and hence responsiveness to reasons.

The discussion of this and the preceding few sections shouldn't cover over an important point. Perry's arguments concern essential indexicals and not essential names or definite descriptions or kind terms. The interpretation of another agent's action as rational does not appear to demand the attribution to that agent of a particular non-indexical belief in place of any other non-indexical beliefs. However, it can demand the attribution of an indexical belief in place of any non-indexical belief. It remains the case that there is a special indexical/non-indexical asymmetry.

8.6 Summary

The argument from action goes beyond issues of cognitive significance because whilst one might try—though I suspect unsatisfactorily—to explain a difference in cognitive significance without reference to a difference in the facts one is aware of one cannot respond to the argument from action in this way.

The argument from action will have implications for all cases in which a rational action requires one belief in place of another. In all such cases the beliefs must involve an awareness of different facts. That the argument is correct to say this about a variety of cases is shown by the fact that definite descriptions pick out different facts even if they denote the same individual. It is also shown by the plausibility of the descriptive theory of names and natural kind terms or, at least, by any theory that accounts for a difference in cognitive significance between co-referring names or kind terms through taking different facts to be associated with the different names or kind terms.

If the role of goal giving desires in rational actions is understood to be simply that the goal adopted is the action performed, then the argument from action also implies that one must adopt a fine grained approach to action individuation. This is an approach with independent plausibility. However, it appears that the goal giving desire must meet further rational requirements. As the state of motivation in acting the desire must also be responsive to the reasons one acts for. Therefore, the argument from action in fact only requires a fine grained approach to desire individuation.

The argument from action is plausible in its implications and numerous applications besides its application to tensed and first-personal beliefs. Therefore, its application to tensed and first-personal beliefs remains appropriate and plausible despite these further implications and applications.

Notes

1. Cf. Frege (1892/1980). I follow Salmon (1986) and others in using this name.
2. Cf. Mill (1884). (However, Mill also uses the term 'name' to refer to a number of expressions besides proper names). This terminology is used by a number of writers including Lycan (2008).
3. A similar puzzle also arises for indexicals because it is possible to find "this is that" informative even if the two demonstratives pick out the same object and despite the thought that it is generally trivial to be told that something is self-identical. (Cf. Perry (1977).)
4. Cf. Russell (1905, 1968) and e.g. Quine (1956).

5. Cf. Perry (1979).
6. I am not sure it makes sense to say an agent must have a belief involving one demonstrative rather than another. Still, there is scope for me to address this case as I address the cases of names and descriptions below. According to many theories of demonstratives an agent must recognize the referent of a demonstrative to be playing a particular role in a context in order to understand the use of the demonstrative. It follows that one will have to recognize different things in understanding different demonstratives. In short, the different demonstratives will involve capturing different facts even if these facts are not part of the truth-evaluable content of the utterances at issue (e.g. the fact that *x* is the most salient thing in context *y* is a different fact from the fact that *x* is the most salient thing in context *z* if the two contexts *y* and *z* are different). E.g., some say you must recognize the referent as the thing demonstrated (e.g. Kaplan (1989a), Reimer (1991a), and perhaps Salmon (2002)), others say that you must recognize it to be the most salient thing in the context (e.g. Wettstein (1984), Reimer (1991b)), or as the object of the speakers intention in the context (e.g. Kaplan (1989b), Bach (1992a, 1992b)), or as the thing in a particular spatiotemporal relation to the speaker in the context (e.g. McGinn (1981)). Thus the argument from action is vindicated in the case of demonstratives.
7. This enables Russell to deal with a number of other puzzles such as how to make sense of non-referring names or non-denoting definite descriptions (cf. Russell (1905)).
8. E.g. Russell (1919, ch. XVI, 1968, lect. II) for 'this' anyway. Perhaps Russell changes his mind about 'I' (e.g. Russell (1910, 1940/1995)).
9. It can be hard to find one description that could be a name or its sense. For this reason some have offered a view according to which names correspond to clusters of definite descriptions (cf. Searle (1958) and Wittgenstein (1953/2001, pt. I, sect. 79)).
10. Russell's theory of definite descriptions has been criticized quite independently from the idea that names are abbreviated definite descriptions, e.g., by Strawson (1950) and Donnellan (1966) who both insist that definite descriptions can refer. However, Kripke (1977) has defended Russell's analysis arguing that it captures the semantic nature of definite descriptions which is distinct from the pragmatic fact that such descriptions can be used by speakers to refer. For more recent discussion see Neal (1990, ch. 3).

11. Cf. Kripke (1981) and also Marcus (1961).
12. There is an awkwardness in applying this argument to certain interpretations of Frege. Because, Frege speaks of the sense of an expression as the mode of presentation of its referent and although he says that this might be captured by a definite description one might wonder whether the object presented is an essential part of the mode of presentation. Evans (1982) has argued that it ought to be thought of as such. I will not discuss this issue further here except to say that even if the referent is taken to be an essential ingredient to the mode of presentation it is still plausible that different modes of presentation give different facts because they give different facts about the referent.
13. Cf. Bach (1981), Loar (1980), and Geurts (1997).
14. Kripke's own theory is that a name refers through a causal link back to a baptism in which the name is given to the referent. This causal link is what is being made use of in descriptions of these types.
15. Jackson (1998b). This is criticized by Soames (2005a, 2005b) for being too powerful, however, Jackson (2007) responds by accepting its potency but denying this undermines it.
16. Cf. Plantinga (1978) who takes the descriptive content to capture the essence of the referent of a name.
17. Kripke (1981, 44) is careful to stress that possible worlds are stipulated not discovered so we do not need to identify a referent in a possible world via its properties. However, this does not really seem to undermine Jackson's point which could be put in terms of our willingness to stipulate a world in which a referent has certain qualities but not one in which it lacks those qualities.
18. Alternatively one could use something like Kaplan's dthat operator (cf. Kaplan (1970)).
19. Russell (1905) was also clearly aware of this distinction of scope and made use of it even if not directly as is required here.
20. Soames (1998) has argued against this use of scope on the grounds that it does not work in particular complex cases, e.g. in some propositional attitude contexts. However, it is not clear that a more complicated rule than the simple 'names take wide scope in modal contexts' cannot be offered to account for such cases. Soames (2005b, 423) later admits this possibility, however, he thinks that the earlier problems associated with K1 will still raise trouble for the descriptivist. Jackson (2007, 26) responds to Soames' argument by suggesting that there are a variety of

complications involved in propositional attitude reports which must be taken into consideration (though Jackson is primarily concerned with descriptions rigidified by 'actually' rather than with talk of scope).

21. I am assuming simple descriptivist theories like Frege's and Russell's. There is of course scope to say that descriptions fix reference or denotation but to insist that they are not to be considered in determining the truth-value of a sentence which is instead determined by the referent/denotation. (Recanati (1993) divides these two roles.)
22. It could even be that the name has different descriptions associated with it by different speakers, though this would put pressure on the notion of meaning (cf. Jackson (1998b, 214–215)).
23. A more detailed discussion of the issues of a prioricity and necessity can be found in discussions of two-dimensional semantics which allow for contingent a priori beliefs and necessary a posteriori beliefs by distinguishing two dimensions of meaning. E.g., one which gives the referent of a term in a possible world considered as counterfactual and one that gives the referent of a term in a possible world considered as actual. (Cf. Davies and Humberstone (1980) and Jackson (1998a, ch. 2).) Jackson utilizes two-dimensional semantics within a descriptive theory of meaning, though the two needn't coincide (e.g. Chalmers (2006) makes it clear that he is not committed to descriptivism by his brand of two-dimensional semantics and Davies and Humberstone (1980), and Kaplan (1989a) do not apply their two-dimensional framework to proper names).
24. E.g. D'Cruz (2000) and Evans (1973).
25. Soames (2002, ch. 2). In a different way see Recanati (1993) who, although he associates a description with a name, does not intend the description to play quite those roles envisioned by Frege and Russell because he takes the description to fix the referent of a name but not to enter into the proposition expressed. Even if the description is removed from the propositional content, as Recanati does, this does not undermine my position because all I require is that an individual is aware of different facts whether or not these fall within the scope of the proposition expressed or the truth-conditions of the utterance.
26. At least this is true for names lacking descriptive content unlike names such as 'Durham University'.
27. Again, my concern is with beliefs rather than with utterances, however, it is plausible to think that if utterances would carry different information

in a particular context, then associated beliefs would too. Soames' talk of different information being conveyed is in part reliant on the epistemic state of the listeners and this is something which appears to be absent in the case of beliefs. However, I think there is a sense in which we place ourselves in a context in talking of another's beliefs and in this way there is scope to find a surrogate for Soames' listener in the belief case. More generally, we can imagine an individual having one belief rather than another just as we can imagine him deliberately using one name rather than another in a situation.

28. E.g. such an idea is explicit in McMullen (1985) and I take it also to be implicit in form in ideas given by Perry himself in more recent work (e.g. Perry (1997, 1998, 2006)). In these works Perry talks about storing information in different files with these files plausibly differing between uses of names and uses of the first-personal indexical even if the names and the indexical are co-referential. Thus my believing 'Carroll is making a mess' will entail me associating the individual making mess with being a philosopher whereas my believing 'Dodgson is making a mess' will not involve me taking the mess maker to be a philosopher but instead to be a mathematician even though the thoughts, unbeknownst to me, concern the same person. These consequences also appear to follow from Alward's (2009) position despite his moving away from Perry's talk of files and instead talking of teams of cognitive relations.
29. It is worth noting that various criticisms of Russell's theory often still presuppose the idea that different definite-description involving beliefs will capture different facts. This is shown, e.g., in the case of Strawson (1950). The use of the definite description to refer presupposes the ability to recognize the referent as fulfilling the criteria stated by the description and hence to recognize that it is the case that, e.g., Carroll is referred to by the name 'Carroll'. Similarly, Donnellan (1966) thinks that although the referent does not need to fulfil the criteria associated with the definite description (language users might be wrong about the world) the referent should nonetheless be taken to fulfil the criteria or be recognized as being taken to.
30. Cf. Chisholm (1970) especially pp. 21–22 for a different argument that they are distinct facts because they entail different counterfactuals.
31. This is clear in many of the works cited in the previous section, however, Soames (2002) does believe that the case of names and kind terms must be treated differently.

32. It is important to note that we are here concerned with a theory of reference/denotation which appears to render the regress or circle vicious. Some people have defended a form of linguistic holism according to which the meaning of each expression is dependent upon the meaning of every other (cf. Wittgenstein (1953/2001, pt. I, sect. 199), Quine (1951), and Davidson (1967/2001b)). But it appears that holism works on the assumption that all words are equal so that a definition tells you as much about the definians as about the definiendum. However, the descriptive theory of reference appears to place words on a hierarchy. You must understand the associated description before, or independently of, understanding the expression whose sense it provides. If I am wrong about this and descriptivism doesn't entail a hierarchical approach to language, then holism might provide a response to the worry I'm outlining in the text here.
33. E.g., Devitt (1996, 159), as discussed by Jackson (1998b). Part of Devitt's point is that if we have to accept a form of reference which is not descriptive, then it is better to apply this to names and natural kind terms too. But, the possibility of necessary a posterior truths and contingent a priori truths, and the explanation of this via two-dimensional semantics, arguably gives a reason for differentiating such cases.
34. The same might also be said for non-natural kind terms too if need be.
35. E.g. Anscombe (1957/2000) and Davidson (1963/2001a).
36. E.g. Goldman (1971) and Thomson (1971a).
37. I will not here go into the issue of what should be taken as the so called basic action, e.g., considering whether it is a mental event, act of the will, contraction of the muscles, etc. I will instead focus on whether pumping and killing, e.g., are the same actions. (Cf. Hornsby (1980, ch. 1 and 2, 1982, 1983), Lowe (1981, 1983), and Alvarez and Hyman (1998).)
38. Cf. Anscombe (1957/2000) and Davidson (1963/2001a).
39. Cf. Goldman (1971), Thomson (1971b), Hornsby (1980, ch.1), and Mackie (1997).
40. Cf. Goldman (1971), Thomson (1971b), and Botting (2010).
41. Cf. Hornsby (1980, ch. 1).
42. Cf. Botting (2010). Anscombe (1979) offers a different response to the multiplier's causal argument, but Botting (2010) gives grounds for thinking this is insufficient (e.g. because it presupposes a certain ontology of events as particulars which is itself questionable).
43. Cf. Anscombe (1979) and Bennett (1973).

44. Cf. Botting (2010).
45. Goldman's (1971) theory of actions would be sufficiently fine grained to fit this notion. However, one must be careful how one accounts for this. For Goldman an action is simply an agent instantiating an action type at a time. It follows that the same action occurs so long as the same agent and action type occur at the same time. For Goldman so long as we see the properties of *being a mechanic* or *being a neighbour* as applied to the agent we are dealing with the same action. However, Goldman does think that action types are very finely individuated so the action type of *a strolling by a mechanic* and the action type of *a strolling by a neighbour*, will be different action types.

References

Alvarez, M., & Hyman, J. (1998). Agents and Their Actions. *Philosophy, 73*, 219–245.
Alward, P. (2009). The Inessential Quasi-Indexical. *Philosophical Studies, 145*, 235–255.
Anscombe, G. E. M. (1979). Under a Description. *Nous, 13*, 219–233.
Anscombe, G. E. M. (1957/2000). *Intention*. London: Harvard University Press.
Bach, K. (1981). What's in a Name? *Australasian Journal of Philosophy, 59*, 371–386.
Bach, K. (1992a). Intentions and Demonstrations. *Analysis, 52*, 140–146.
Bach, K. (1992b). Paving the Road to Reference. *Philosophical Studies, 67*, 295–300.
Bennett, J. (1973). Shooting, Killing, and Dying. *Canadian Journal of Philosophy, 2*, 315–323.
Botting, D. (2010). Three Theses on Acts. *Philosophical Explorations, 13*, 65–79.
Chalmers, D. J. (2004). Epistemic Two-Dimensional Semantics. *Philosophical Studies, 118*, 153–226.
Chalmers, D. J. (2006). Two-Dimensional Semantics. In E. Lepore & B. C. Smith (Eds.), *The Oxford Handbook of Philosophy of Language* (pp. 574–606). Oxford: Oxford University Press.
Chisholm, R. (1970). Events and Propositions. *Nous, 4*, 15–24.
Davidson, D. (1963/2001a). Actions, Reasons, and Causes. In *Essays on Actions and Events* (pp. 3–19). Oxford: Oxford University Press.

Davidson, D. (1967/2001b). Truth and Meaning. In *Inquiries into Truth and Interpretation* (pp. 17–36). Oxford: Oxford University Press.
Davies, M., & Humberstone, L. (1980). Two Notions of Necessity. *Philosophical Studies, 38*, 1–31.
D'Cruz, M. (2000). A Theory of Ordinary Proper Names. *Mind, 109*, 721–756.
Devitt, M. (1996). *Coming to Our Senses: A Naturalistic Program for Semantic Localism*. Cambridge: Cambridge University Press.
Donnellan, K. S. (1966). Reference and Definite Descriptions. *Philosophical Review, 75*, 281–304.
Dummett, M. A. E. (1973). *Frege: Philosophy of Language*. London: Duckworth.
Evans, G. (1973). The Causal Theory of Names. *Proceedings of the Aristotelian Society, Supplementary Volume, 47*, 187–208.
Evans, G. (1982). In J. McDowell (Ed.), *The Varieties of Reference*. Oxford: Oxford University Press.
Frege, G. (1892/1980). On Sense and Meaning (M. Black, Trans.). In P. Geach & M. Black (Eds.), *Translations from the Philosophical Writings of Gottlob Frege* (pp. 56–78). Oxford: Blackwell.
Geurts, B. (1997). Good News about the Description Theory of Names. *Journal of Semantics, 14*, 319–348.
Goldman, A. I. (1971). The Individuation of Action. *The Journal of Philosophy, 68*, 761–774.
Hornsby, J. (1980). *Actions*. London: Routledge and Kegan Paul.
Hornsby, J. (1982). Reply to Lowe on Actions. *Analysis, 42*, 152–153.
Hornsby, J. (1983). Events that are Causings: A Response to Lowe. *Analysis, 43*, 141–142.
Jackson, F. (1998a). *From Metaphysics to Ethics: A Defence of Conceptual Analysis*. Oxford: Oxford University Press.
Jackson, F. (1998b). Reference and Description Revisited. *Philosophical Perspectives, 12*, 201–218.
Jackson, F. (2007). Reference and Description from the Descriptivist's Corner. *Philosophical Books, 48*, 17–26.
Kaplan, D. (1970). Dthat. *Syntax and Semantics, 9*, 221–243.
Kaplan, D. (1989a). Demonstratives. In J. Almog, J. Perry, & H. Wettstein (Eds.), *Themes from Kaplan* (pp. 481–563). Oxford: Oxford University Press.
Kaplan, D. (1989b). Afterthoughts. In J. Almog, J. Perry, & H. Wettstein (Eds.), *Themes from Kaplan* (pp. 565–614). Oxford: Oxford University Press.
Kripke, S. (1977). Speaker Reference and Semantic Reference. *Midwest Studies in Philosophy, 2*, 255–276.

Kripke, S. (1981). *Naming and Necessity*. Oxford: Blackwell.
Loar, B. (1980). Names and Descriptions: A Reply to Michael Devitt. *Philosophical Studies, 38*, 85–89.
Lowe, E. J. (1981). All Actions Occur Inside the Body. *Analysis, 41*, 126–129.
Lowe, E. J. (1983). Reply to Hornsby on Actions. *Analysis, 43*, 140–141.
Lycan, W. G. (2008). *The Philosophy of Language: A Contemporary Introduction*. Abingdon: Routledge.
Mackie, D. (1997). The Individuation of Actions. *Philosophical Quarterly, 47*, 38–54.
Marcus, R. B. (1961). Modalities and Intensional Languages. *Synthese, 13*, 303–322.
McGinn, C. (1981). The Mechanisms of Reference. *Synthese, 49*, 157–186.
McMullen, C. (1985). 'Knowing What It's Like' and the Essential Indexical. *Philosophical Studies, 48*, 211–233.
Mill, J. S. (1884). *A System of Logic, Ratiocinative and Inductive*. London: Longmans.
Neal, S. (1990). *Descriptions*. Cambridge, MA: Massachusetts Institute of Technology Press.
Perry, J. (1977). Frege on Demonstratives. *Philosophical Review, 86*, 474–497.
Perry, J. (1979). The Problem of the Essential Indexical. *Nous, 13*, 3–21.
Perry, J. (1997). Indexicals and Demonstratives. In B. Hale & C. Wright (Eds.), *A Companion to the Philosophy of Language* (pp. 586–612). Oxford: Blackwell.
Perry, J. (1998). Myself and I. In M. Stamm (Ed.), *Philosophie in Synthetisher Absicht (A festschrift for Dieter Heinrich)* (pp. 83–103). Stuttgart: Klett-Cotta.
Perry, J. (2006). Self. In *Encyclopedia of Philosophy* (Vol. 8, pp. 708–711). Detroit: Macmillan.
Plantinga, A. (1978). The Boethian Compromise. *American Philosophical Quarterly, 15*, 129–138.
Quine, W. v. O. (1951). Two Dogmas of Empiricism. *The Philosophical Review, 60*, 20–43.
Quine, W. v. O. (1956). Quantifiers and Propositional Attitudes. *Journal of Philosophy, 53*, 177–187.
Recanati, F. (1993). *Direct Reference: From Language to Thought*. Oxford: Blackwell.
Reimer, M. (1991a). Demonstratives, Demonstrations, and Demonstrata. *Philosophical Studies, 63*, 187–202.

Reimer, M. (1991b). Do Demonstrations Have Semantic Significance? *Analysis, 51*, 177–183.
Russell, B. (1905). On Denoting. *Mind, 14*, 479–493.
Russell, B. (1910). Knowledge by Acquaintance and Knowledge by Description. *Proceedings of the Aristotelian Society, 11*, 108–128.
Russell, B. (1919). *Introduction to Mathematical Philosophy*. London: Allen & Unwin.
Russell, B. (1940/1995). Egocentric Particulars. In *An Inquiry into Meaning and Truth* (pp. 108–115). London: Routledge.
Russell, B. (1968). The Philosophy of Logical Atomism. In R. C. Marsh (Ed.), *Logic and Knowledge: Essays 1901–1950* (pp. 175–282). London: Allen & Unwin.
Salmon, N. V. (1986). *Frege's Puzzle*. Cambridge, MA: Massachusetts Institute of Technology Press.
Salmon, N. V. (2002). Demonstrating and Necessity. *The Philosophical Review, 111*, 497–537.
Searle, J. R. (1958). Proper Names. *Mind, 67*, 166–173.
Soames, S. (1998). The Modal Argument: Wide Scope and Rigidified Descriptions. *Nous, 32*, 1–22.
Soames, S. (2002). *Beyond Rigidity: The Unfinished Semantic Agenda of Naming and Necessity*. Oxford: Oxford University Press.
Soames, S. (2005a). *Reference and Description: The Case against Two-Dimensionalism*. Oxford: Princeton University Press.
Soames, S. (2005b). Reference and Description. In F. Jackson & M. Smith (Eds.), *The Oxford Handbook of Contemporary Philosophy* (pp. 397–426). Oxford: Oxford University Press.
Strawson, P. F. (1950). On Referring. *Mind, 59*, 320–344.
Thomson, J. J. (1971a). The Time of a Killing. *Journal of Philosophy, 68*, 115–132.
Thomson, J. J. (1971b). Individuating Actions. *Journal of Philosophy, 68*, 774–781.
Wettstein, H. (1984). How to Bridge the Gap Between Meaning and Reference. *Synthese, 84*, 63–84.
Wittgenstein, L. (1953/2001). *Philosophical Investigations* (G. E. M. Anscombe, Trans.). Oxford: Blackwell.

9

The Argument from Appropriate Emotions

In Chap. 7 I argued that the tenseless theory of time was incorrect because the demand for tensed beliefs in rational actions proved that those beliefs captured facts not captured by any tenseless beliefs. This argument rested on the argument for the essential indexical and on the nature of rational action.

In this chapter I will examine the nature of emotions, or a sub-species thereof, and their appropriateness. I will defend a cognitive view of emotions according to which an emotion involves a belief. I will argue that whether such emotions are appropriate or inappropriate is determined by the reasons that those emotions are felt for. The cognitive component of an emotion will be an awareness of the reasons that emotion is felt for.

I will take up issues raised in Chap. 2. On the basis of Prior's thank goodness argument I will show that an emotion can be appropriate when it is felt for a reason captured by a tensed belief and inappropriate when felt for a reason captured by any tenseless belief. This will enable me to argue, in a manner analogous to Chap. 7, that the tensed belief must capture a reason—or fact—not captured by any tenseless belief.

Olley (F.O.C.H.) Pearson, *Rationality, Time, and Self*,
https://doi.org/10.1007/978-3-319-71973-3_9

9.1 A Cognitive Theory of Emotions

James (1884) said that emotions were the feelings of bodily changes which resulted from a perception of an exciting fact. For example, when I see a bull charge towards me this causes my heart rate to rise and adrenalin to flow, etcetera. I will feel these physiological changes and these feelings will be my emotion of fear.

However, this view that emotions are simple[1] feelings suffers from a number of criticisms. Cannon (1917) has argued that physiological data does not support James's claim because one can have the physiological effects and not the emotion. For example, one can have the physiological effects associated with fear, such as a rising heart rate, and not have the emotion of fear.

Separating the feeling and a particular physiology does not help the view that emotions are simple feelings. Feelings are not fine grained enough to distinguish the myriad of emotions that people have. For example, feelings of embarrassment and of shame can coincide. Furthermore, emotions can go unfelt. A person might not realize that she was angry on a past occasion until she looks back and reconsiders it. Alternatively, a person might have an emotion, such as an anger at a childhood slight, for a number of years but only have feelings of that emotion during particular episodes within that time.[2]

It is little better to adopt a form of behaviourism[3] and instead identify emotions with actions or dispositions to act. Behaviourism here suffers many of the criticisms of behaviourism about other mental states. It is not clear exactly what behaviour ought to be associated with a particular emotion. People can feel an emotion but refrain from acting in typical ways because of certain other beliefs and desires that they have. For example, a person may not want to show his or her anger. Moreover, it is clear that people do not generally know about their emotions through observing their own behaviour either directly or indirectly. It is plausible that there is a link between emotions and behaviour. But it is implausible that emotions simply are behaviour.[4]

Solomon[5] points out that people can know what emotion they would likely feel in a particular scenario because of the details of the scenario and the judgments they know they would make about it, and not because they know how they would feel or behave. For example, if I know that I would take a scenario to be awkward but not my responsibility, then I will expect myself to be embarrassed but not ashamed. One can know about one's emotions through knowledge of what one is or would be aware of.

We saw in Chap. 2 that MacBeath criticized Mellor for failing to acknowledge that people are emotional about things. For example, Prior is relieved that the exams are over, Mellor is grateful that MacBeath offered him an improved solution to Prior's puzzle, and so on. It is clear that emotions have cognitive content. They involve a state of awareness, or belief.[6] This shows that emotions are not simple feelings. It also enables one to account for why there would be a link between particular emotions and particular types of behaviour. It is because there is a link between particular beliefs and particular types of behaviour.

Recognizing that emotions have a cognitive component also nicely fits the fact that there are fine-grained distinctions between types of emotions, because there are fine-grained distinctions amongst types of beliefs. Moreover, there are plausible identifications between types of the former and types of the latter sort. For example, in fear one will believe something to be dangerous, in relief one will believe a discomfort has ended, in shame one will judge a discomfort to be one's responsibility, and so on.[7]

This view is also supported by considering uses of emotion language. For example, when an individual says "I am very angry with you" we are informed of that individual's emotion. But, very often the appropriate way to respond to this is not to react as if we have been informed of another's feelings. For example, it is not appropriate to respond by saying "Poor you, that must be uncomfortable". It is also not appropriate to react as if we have been informed of another's behaviour. We shouldn't say "Thanks, in that case I'll avoid you for a while". Rather, we ought to respond to the utterance as informing us of a belief the speaker has about us, such as the belief that we have done something wrong. We ought to say "I'm sorry, it was an accident".[8]

Therefore, emotions are not simple feelings or patterns of behaviour. Emotions have a cognitive component. In having an emotion a person believes something to be the case. This is the *cognitive theory of emotions.*

9.2 Objections to a Cognitive Theory of Emotions Rebutted

The cognitive theory of emotions has increased in popularity. But, the view has also been criticized. I will now consider some of these criticisms to clarify the view I am proposing.

Solomon (1993) once defended the view that an emotion could be identified with its cognitive component so that feelings were mere ornamentations to emotions but no proper part of them. However, if this view is correct it is not clear how there can be a difference between having an emotion and having a belief with the same content without having the emotion. But this seems possible. For example, I can dispassionately be aware that someone has died without grieving that she has died. But plausibly a belief that someone has died is a paradigmatic example of the cognitive component of grief.[9]

The obvious, and I believe correct, response to this criticism is to accept that emotions are not merely cognitive but are more complex.[10] For the reasons given in the last section it is wrong to say simply that an emotion must be a belief and a particular feeling or belief and a particular pattern of behaviour. Nevertheless, it appears that an emotion will have close links to or be in part constituted by feelings. Plausibly it will also have close links to certain forms of behaviour and facial expressions. I will not try to detail this relation. But the cognitive view of emotions should not be understood to be that an emotion is no more than a belief: not all states of awareness qualify as emotions.[11]

It appears that someone might be afraid and not believe that he or she is in danger and therefore it is wrong to take an emotion of fear to involve a belief that one is in danger. For example, someone on a high balcony might appear to be scared even though he says that he believes that he is not in danger because he believes the rail is strong, and no one will push him, etcetera.[12] This is an example of a *recalcitrant emotion*.

The force of this criticism against the cognitive theory is removed by the peculiarity of the situation at hand. Most often when people are fearful if they come to believe that they are not in danger, then they will lose their fear. For example, if I step out onto a road to cross it and glimpse a car out of the corner of my eye, then I will become afraid. But when I look more closely and see that the car is stationary the fear will pass and I will continue in my stride. We should therefore expect there to be something peculiar in a situation in which someone professes to know he is in no danger and yet appears to be fearful. We are encouraged not to treat the case simply at face value or hence to see it as a straight forward counter example to the cognitive theory.

For example, it might be that the individual accepts that there is good evidence that he is safe and he does not want to appear sceptical of his friend's assertions that he is safe, and therefore he says that he believes that he is safe. Nevertheless, the individual does actually harbour a belief that he is in danger. He passed the building when it was in construction and saw the workmen hurrying. He knows building inspectors can be susceptible to bribes even if this isn't very common and so on. Perhaps the individual would admit that in fact he believes there is a chance he will fall if pressed on the issue. But even if that is not the case the appearance of fear, his unwilling to approach the edge, and the peculiarity of the situation are already evidence that such a belief might be at play. Alternatively, it might be that the person really does believe himself to be safe and that he is not in fact scared. Perhaps he is simply experiencing physiological effects related to fear or a feeling related to fear as the result of some sort of pathology. A pathology which is indicated by the peculiarity of the case. Finally, it could be that the person holds contradictory beliefs because he both takes himself to be in danger and takes himself not to be in danger. Accusing an individual of self-contradiction is not something that should or even could often be appropriate. But, as already noted, the case at issue is a rare one and, moreover, it is a case in which the individual concerned appears to be irrational. (I will say more about recalcitrant emotions in Sect. 8.4 below.)

The idea that emotions are cognitive has also been criticized on the grounds that animals and infants can have emotions but it is not clear that they can have complex cognitive states. For example, it appears animals and infants can be afraid but it is not so clear that they can believe that they are in danger. The problem is that 'danger' is a complex concept and therefore not one we would readily attribute to animals and infants.[13]

However, it is quite plausible that animals and infants do have a less rich emotional life than more mature humans.[14] Furthermore, it is plausible that to the degree that animals and infants have emotions they have them about something. Their emotions are cognitive. The cat is afraid of the dog. It might be that we ought therefore to say that fear doesn't simply involve the belief that one is in danger. Fear could also involve a more simple belief such as 'pain is imminent'.[15] Alternatively, it might be that we ought to say that fear concerns danger and that animals and infants have a variety of emotions which are related to and perhaps predecessors of fear. These related emotions might be more simple and involve more simple beliefs, such as 'that causes pain'. I will not decide between these options but on either account it remains the case that emotions are cognitive states and that, for example, fear remains a concern with harm.

Robinson (1995) has criticized the cognitive view of emotions through a consideration of the startle response. Robinson believes that startle is an example of an emotion. It gives rise to changes in facial expressions which is a trait linked with emotions. It is also related to fear and shock which are characteristic emotions. Startle, however, is not cognitive. It doesn't involve the agent having a belief about what startled him or her. It appears to follow that emotions needn't be cognitive.

I believe that startle is not an emotion[16] but is rather a more primitive brute reaction to the environment. Nevertheless, I will not defend this idea. I will instead limit my discussion to emotions that are more obviously cognitive. I will not define this group of emotions but clear examples include the following: relief, gratitude, anxiety, grief, anger, fear, embarrassment, contempt, and shame. These examples are all linked in a manner similar to that which Anscombe (1957/2000, ch. 2) links all actions. They are all emotions for which people can give reasons as to why they have them or why someone would be justified in having them. They are emotions that inform us of the beliefs and values of the person who has the emotion. If startle is an emotion and it is not cognitive, that does not entail that all emotions ought to be thought of in that manner and nor does it undermine the points raised above in favour of thinking of some emotions as cognitive.[17] I shall be concerned with just this class of cognitive emotions and I will simply use the term 'emotion' to refer to members of this class unless otherwise indicated.

The cognitive theory of emotions that I propose is therefore not the view that all emotions are simply beliefs. Rather, by 'emotions' I mean to pick out a class of cognitive emotions which may or may not be the class of all emotions. Further, these emotions will be something more than mere beliefs insofar as they involve or are constituted by feelings and are linked with types of behaviour.

9.3 The Appropriateness of Emotions

For some time it was thought that emotions and rationality were entirely distinct. Hume (1739–40/1985, bk. II, pt. III, sect. III), for example, declared that reason will always be the slave of the emotions, because the emotions could not be compelled or judged by the standards of rationality. However, more recently[18] room has been made to consider the rationality or appropriateness of emotions. There are two broad forms that this has taken in the literature. One of these considers the practical value of an emotion and one considers the rationality of an emotion's cognitive component.[19]

That emotions can be rational or justified is evinced by our everyday practices. For example, it is inappropriate to get angry at someone for being nice to you or to be extremely grateful to someone for being just a little bit nice to you. In fact, that emotions can be appropriate or rational has been offered as evidence that emotions are not simply feelings because simple feelings, such as headaches or itches, are not rational or appropriate (even if they can be expected effects of certain events).[20]

Emotions have been judged to be rational or appropriate in regard to their practical value. Damasio (1995) has argued that emotions play a vital role in reasoning. It is thought that emotions can provide a solution to the frame-problem by making certain facts salient to an agent in his or her reasoning and therefore preventing the agent from being drawn into a potentially endless consideration of factors. Emotions are said to have a rational value because they enable agents to act rationally without being drawn into endless reasoning.

However, it is not clear that emotions do play this practical role. Evans (2002), for example, has argued that before we can be sure that emotions provide a solution to the frame-problem we need a clear account of what emotions are and how they play this role. Moreover, even if emotions do play this role it is not clear that we are not passing over a different and better method for solving the problem in letting them do so. Therefore, it is not clear that we should take emotions to be appropriate or rational because they fulfil this role. The emotions might block better ways and so be a hindrance.

Elster (2004) suggests a different manner in which emotions have practical value: they are the sole source for the maintenance of social norms. Social norms are non-outcome-orientated injunctions to behave in certain ways. They are distinct from social conventions. An example of the former is the norm of wearing black at funerals and an example of the latter is the convention of driving on the left in the UK. According to this view my shame in my not wearing black at a funeral will be a response to another's contempt for me not wearing black and will in turn bring me to feel contempt for others who break the norm.[21] These emotions ground the norm.

But Elster's suggestion is implausible. It is not clear what social norms could be in play in some cases of appropriate emotions. For example, no social norm appears to be supported by my being relieved that my headache has ended and it is not the case that I ought to be ashamed if I am not relieved it has ended or that others ought to experience contempt towards me because I am not relieved. I have not worn bright colours to a funeral. But this is not because I have experienced contempt as a result of doing so (I haven't done so). It is because I was told as a child that I ought not to. In short, there are clearly other ways than the experience of emotions to pass on social norms. Furthermore, if emotions did play the role of maintaining social norms, this would only make the emotions appropriate if the social norms themselves were appropriate. But it is plausible that if the norms are appropriate, then the appropriateness of the emotions stems from them and not from the role of emotions in maintaining them. If I am ashamed at having worn bright colours to a funeral, I will not be ashamed because of having experienced contempt. I will not be ashamed of experiencing contempt. I will be ashamed of my wearing the bright clothes. My act, not the contempt, would be the object of my emotion. If the norm of not wearing bright colours is appropriate, then plausibly my act of wearing them is inappropriate and it is my shame's having this act as object that makes it appropriate (whether or not it was triggered by contempt).

The practical value of emotions needn't rest solely on general claims such as those concerning the frame-problem or social norms. More specific claims might be made instead. For example, the discomfort of anticipating a pain might encourage an agent to act so as to prevent the pain coming around and in turn be an aid to the survival of the agent.[22]

But there are a number of general problems with taking the rationality or appropriateness of an emotion to rest on its practical value. It is not clear that emotions are always of practical value.[23] For example, in a case such as the anticipation one it might well be that anticipation makes an agent nervous and therefore perform less well. So anticipation actually places the agent at greater risk. Further, it is not clear that an emotion such as grief plays any practical role at all. After all, a concern for death or loss is already to be found in emotions such as fear or in sensations such as pain.[24]

Moreover, these points make it clear that we are not generally certain of whether a particular emotion, or even emotions in general, will be of practical value to an agent. However, we are generally certain of whether an emotion is appropriate or not. Our everyday judgements of the appropriateness or rationality of emotions therefore cannot be based on their practical value. If I do not know whether or not grief is advantageous, my certainty that it is inappropriate to grieve a death that is yet to come cannot be a judgment of the practical value of the emotion. Whether or not emotions are practical is hence beside the point.[25]

Cockburn (1997, 1998, ch. 2) has noted that if someone is angry because she believes that someone has slighted her, then this anger is inappropriate if in fact the person has done no such thing. Similarly, if someone is extremely angry at another who has only done her a very minor injustice, then this anger is inappropriate. These examples of anger remain inappropriate even if it turns out that they were practically fortuitous to the emotional individuals. A celebrity might make her living from being an overly angry character, but that does not mean that her exaggerated emotions are appropriate. These points lend weight to the idea that the appropriateness at issue is not a matter of practical value.

Cockburn presses this idea further by pointing out that there is something worrying, I would say sinister, in the idea that our emotions are purely or ultimately to be valued in terms of their practical worth.[26] It is surely not right to grieve a loved one simply because that brings practical advantages to one. On the contrary, such an emotion would appear highly inappropriate and disingenuous.

The cognitive element of emotions raises scope for a rather different manner in which emotions can be evaluated as appropriate or not. This alternative recognizes that emotions themselves are responsive to rational concerns not merely characters within them. (Hume (1739–40/1985, bk. II, pt. III, sect. III) denied that emotions had a cognitive component and therefore this approach could not have been open to him.) Beliefs can be appropriate or inappropriate and therefore if emotions involve beliefs, this evaluation should also apply to them. The appropriateness of a belief is grounded in its origin. An appropriate belief will have an appropriate origin. An appropriate origin is one that is likely to entail the truth of the belief.[27] For example, if I believe that it is raining because I witness the rain, then my belief is appropriate. Whereas, if I believe that it is raining because I dream that it is, my belief is not appropriate. Emotions are thus evaluated as appropriate or not according to whether the beliefs they involve are appropriate or not. If I am angry because someone has slighted me, I will believe that someone has slighted me. If this belief would be appropriate, then, so the theory goes, the anger would be too.

It is plausible that we can speak of the appropriateness of emotions in terms of the appropriateness of the beliefs they involve. However, it is wrong to think that this is the only or fundamental manner in which we so speak, because emotions can be appropriate or not independently of whether or not the beliefs they involve are appropriate. If I learn that a loved one will die on a particular date in the distant future, it is inappropriate for me to at that time grieve the death. That is, it is inappropriate to have grief where the belief element involved is the belief '*x* will die at *t*' or the tenseless belief '*x* dies at *t*'. Such grief would be inappropriate despite the belief it involves being true and appropriate (that is, originating in a source likely to lead to true beliefs). Similarly, if I am in danger it is appropriate to be scared, but it is not appropriate to be scared of that danger afterwards. That is, it is appropriate to have fear involving the belief 'I could fall off this balcony which would cause me great harm' but not involving the belief 'I could have fallen of that balcony which would have caused me great harm' despite both beliefs being true and appropriately formed. Alternatively, if my friend wrongs me in a minor way it is appropriate for me to be a little bit angry, but it is not appropriate for me to be extremely angry. But plausibly both emotions would involve the same belief, namely, 'my friends refusing my invitation was a slight against me'. Some things Pitcher (1965) says suggest the thought that extreme anger and mild anger will involve different beliefs, for example, 'I was badly wronged' and 'I was slightly wronged' respectively. But I find this implausible. It is clear people can get extremely angry about what they know is a trivial matter. Moreover, their anger can calm and nothing about the belief involved changes in this calming.

The appropriateness of emotions therefore does not reduce simply to the appropriateness of the beliefs they involve.[28] It is not simply a matter of whether the beliefs involved originate in a source likely to give rise to true beliefs. But this perhaps ought to be expected. It is fitting that there is something more to the appropriateness of emotions because emotions are more than just the beliefs they involve. In short, it is fitting that there is something that makes it appropriate to have an emotion on an occasion not merely to have a simple belief which would capture its cognitive component.

As already intimated above, emotions are had for reasons. I am angry because someone has wronged me.[29] I give her wronging me as a reason for me to be angry and her wronging me makes it reasonable for me to be angry. If an emotion is had for a reason that is a reason that justifies that emotion, then that emotion will be appropriate. It will be reasonable. In the terminology of Chap. 4 we might say that if the m-reason for my emotion is a j-reason for that emotion, then the emotion is appropriate. This is a form of appropriateness distinct from either the practical value of the emotion or the appropriateness of the belief involved as just discussed. Furthermore, this form of appropriateness is fit to fulfil the roles unfulfilled by the previous forms. For example, it is appropriate of me to be angry when my friend slights me by refusing my invitation because this is a j-reason for my being angry. However, this is not a j-reason to be extremely angry and therefore extreme anger would be inappropriate.

Thus we can see that there are number of different manners in which one might judge an emotion to be appropriate. Recent literature's two most popular types of way of doing this—in respect of the practical value of the emotion and in respect of the appropriateness of the belief involved in the emotion—cannot exhaust the field. We must recognize a third measure, namely, whether the emotions are had for reasons that justify those emotions. Moreover, this third measure best captures our general and everyday evaluations of emotions. We readily judge that when a person is extremely angry for the reason that she has suffered a very minor injustice this emotion is inappropriate. We make this judgement despite recognizing that the emotional individual gained the belief that she had suffered a minor injustice from an appropriate source and being completely unaware whether the emotion was of any practical value.

9.4 The Reasonableness of Emotions

In the previous section we saw that emotions are usually identified as appropriate or not simply in regard to the reasons that they are had for. In this section I will consider what these reasons are and how they are related to the emotions.

It is clear that the reason an emotion is had for, its m-reason, is given through the cognitive component of the emotion. When I feel relief because—for the reason that—my headache is over the cognitive component of my relief will be the belief 'my headache is over'. Similarly, when I am afraid because I might fall off the balcony which would cause me harm the cognitive component of my emotion will be the belief 'I might fall off this balcony which would cause me harm', and so on. Conversely, when I have grief whose cognitive component is the belief 'my grandmother has died' the reason for which I grieve is that my grandmother has died and so on.

Solomon (1993, ch. 5) has suggested that the cognitive component of an emotion picks out a subjective object which is something other than a fact or a state of the world. For Solomon, when I am relieved because my headache is over my relief does not concern the fact that my headache is over. Rather, it concerns a subjective object.[30] Moreover, this subjective object can exist even if the fact does not, that is, even if my headache is not over.

However, it is clear that Solomon is wrong. If my headache is not over I ought not to be relieved that it is. Furthermore, if I know that my headache is not over I will not be relieved that it is. I do not take this subjective object to be a j-reason for my emotion and therefore I do not adopt this subjective object as an m-reason. (The arguments here mirror arguments of Chap. 4, though I take Solomon's subjective object to be something different from a mental state.)

Facts are j-reasons for emotions and subjective objects are not. It is the world that one is responding to that makes emotions justified or appropriate. I ought only to be relieved that my headache is over if it is, to grieve a death if there has been a death, fear a danger if there is a danger, and so on. Further, we are aware of this and therefore we often have these j-reasons as m-reasons. If I learn there was no death, I will stop grieving (even if you tell me a subjective object still exists). If I learn there is no danger, I will stop feeling scared (even if you tell me a subjective object continues to exist), and so on. Moreover, it must be the case that some emotions are had for reasons that justify them and therefore that sometimes our m-reasons for our emotions are j-reasons for them and are hence facts. We do not deliberate about what emotions to feel in the same way that we deliberate how to act. But, we do take a consideration of facts to alter our emotions. We try to make people less angry by telling them facts about how things will turn out okay and it is when we consider the facts of a situation that we become emotional about it. Our everyday interactions evince the idea that we take emotions to be responsive to facts.

Solomon (1993, ch. 5) recognizes that it appears that the cognitive component picks out facts, but he motivates the idea that it picks out subjective objects by considering error cases. For example, when I mistakenly believe that I am in danger and am resultantly afraid. However, with the arguments of Chap. 4 in mind we can see our way clear of this muddle. If I mistakenly believe that I am in danger and this belief is the cognitive component of an emotion of fear, then it does not follow that my m-reason for my fear was some subjective object. Rather, it could simply be that I had no m-reason for my emotion. Furthermore, it appears that in such a case I will not have an m-reason, because if I learn that I am not in danger I will believe that I ought not to have been afraid. I will admit that I had no reason to be scared and merely mistakenly believed that I did. Prior to learning of this mistake I would cite a non-existent—that is, nothing—as my reason and we ought to take this at face value. The emotion is an attitude towards a non-existent—my danger. This is not in itself unusual because we have all sorts of attitudes towards non-existents. Moreover, it reveals that the m-reason is a non-existent, that is, that I have no m-reason.

The arguments of the last paragraph do not undermine the fact that, for example, if one has been misled to mistakenly believe one has lost a loved one and thus to grieve, then that grief might be excusable. However, it remains the case that the belief is false and the emotion is inappropriate. Moreover, that grief is somehow excusable in such an error case derives from the fact that were one not misled so that the belief involved were true, then one would be justified in grieving. In a non-error case the reason makes the emotion appropriate. In the error case there is no reason and the emotion is inappropriate, but the emotion bears a similarity to the non-error case which can make the emotion excusable.

So far the case with emotions is analogous to that with actions. We act for reasons which are facts and which we are aware of through beliefs. However, there is a slight difference. The emotion might be in part constituted by this belief because the emotion is itself a state of awareness.

Error cases do not undermine this. But, it is of note that a rather different challenge to the idea that emotions are responsive to facts can be uncovered in discussions of recalcitrant emotions. Recalcitrant cases, as noted above, have been taken as a challenge to the cognitive view of emotions. The criticism concerns the idea that someone might, for example, appear to be afraid of falling even though she appears to believe that she cannot fall and is in no danger. So we seem to be left with two options. First, the person does not believe she is in danger. Second, the person does believe she is in danger and therefore has contradictory beliefs. The first option conflicts with the cognitive view of emotions, but the second looks unappealing as we do not like to think people are that irrational.

One common form of response to the recalcitrant cases is to suggest that whilst the cognitive component of an emotion concerns facts there is a sense in which these facts do not need to be taken to obtain.[31] This line of response to the recalcitrant case thus conflicts with the datum that emotions are had for reasons that are facts.

For example, when considering recalcitrant cases Greenspan (1988, ch. 1 and 2) says that the cognitive component of the emotion is like imagination. The person imagines that she could fall and is in danger. This imagination can persist even if she is not in danger and even if she believes that she cannot fall and is not in danger. There is nothing irrational in imagining things to be one way and believing them to be another. So there is no problem. The emotion has a cognitive component which is imagination. But, the person only has contradictory beliefs insofar as she imagines things to be one way whilst knowing them to be another and there is nothing worryingly irrational about that.

But, now it is clear that the cognitive component of the emotion serves to pick out the reason the emotion is felt for it is also clear that one should not try to respond to recalcitrant cases merely by weakening the cognitive component. The agent is supposed to merely imagine she might fall and is in danger. But, this imagining is still supposed to provide the agent with her reason for being afraid. If it does not involve taking the danger to be real it does not involve taking there to be a reason to be afraid. This leaves us with difficulties because it now appears the person both does not take herself to have a reason to be afraid—because the cognitive component is mere imagining—and does take herself to have a reason to be afraid—because she is afraid of falling. This line of response to recalcitrant cases therefore fails to show that the emotional individual avoids irrationality because it leaves the emotional individual both taking and not taking herself to have a reason to be afraid.

This inadequacy is revealed if we imagine an analogous case of acting for a reason. We see a person running and ask them why. The person responds by saying he is running in order to catch the bus. We then point out to him that there is no bus. If he replied "I know there is no bus, I am just imagining there is one" then he would strike us as irrational. Softening the awareness of one's reasons to a point that does not involve taking them to obtain does not remove irrationality. It establishes irrationality.

But there is a way to avoid this irrationality. No doubt the individual with recalcitrant fear will say that she does not take her possibly falling to justify or motivate her fear because she says that she does not believe that she could possibly fall. (There will plausibly also be differences in the phenomenology of normal and recalcitrant cases.) We therefore have grounds to doubt that the individual in the recalcitrant case is afraid for a reason.[32] If the individual does not take her possibly falling to be a reason to fear, then she does not need to assent to the content of the cognitive component of her emotion or strictly to believe that she could fall. The self-contradiction and hence irrationality is thereby removed.

If the individual in the recalcitrant case is not afraid for a reason, then in this case the emotion might lack a cognitive component or have a weak one that doesn't involve taking a fact to obtain. If this is the case, it does not go hand in hand with denying that emotions in general have a strong cognitive component, rather quite the opposite. We have seen that normally emotions are had for reasons and this requires assenting to the content of the cognitive component of an emotion. Thus the cognitive component of a normal emotion is taken to be a belief. In the case of a recalcitrant emotion the emotion appears not to be had for a reason and therefore assent and belief are not present. If we remove or weaken the cognitive component in the recalcitrant case this is therefore necessarily just the removal or weakening of the cognitive component of an unusual subset of emotions. This enables one to say that the case of recalcitrant emotions is not a problem for the cognitive theory because it is an unusual case. Moreover, when this is said it does not merely mean that we do not need to worry about outliers. It specifically means that recalcitrant emotions are different in kind from normal emotions because of their unresponsiveness to reasons. Therefore, we shouldn't conclude normal emotions involve anything less than taking facts to obtain.[33]

Recalcitrant emotions therefore do not involve worrying self-contradiction either in terms of the holding of contradictory beliefs or in terms of the taking a reason both to and not to obtain. To this extent the irrationality of recalcitrant emotions is removed. This might worry some, such as Brady, who believe recalcitrant cases do involve some form of irrationality (Brady 2009). However, as Brady himself notes, self-contradiction is not the only possible form of irrationality. Rather, a lack of responsiveness to reasons can render something irrational.[34] For example, adopting a belief simply at whim rather than in response to reasons renders that belief irrational. In this way recalcitrant emotions remain irrational specifically because they are not responsive to reasons as normal emotions are.

The j-reasons for emotions will often be j-reasons to have the beliefs involved in the emotions. For example, the fact that my headache is over is a j-reason for me to believe that my headache is over as well as being a j-reason for me to be relieved that my headache is over. Nevertheless, it remains the case that the j-reason for an emotion is not simply the j-reason for the belief involved in the emotion because, as already noted, the j-reason for the belief involved might not be a j-reason for the emotion. For example, this will be the case if the m-reason for an emotion is not a j-reason to have that emotion as when I am extremely angry at a slight injustice. Before closing this section, I will briefly consider some difficulties for the view that the j-reasons for emotions are facts.

I have spoken of the j-reason for fear being the fact that there is danger and in more detailed moments I have spoken of the fact that it is possible one will be harmed. So my view may imply that there are facts of possibility. This might sound worrying but it is not. It is clear that there can be a fact of the matter whether or not something is possible. But, it might be less clear what the nature of this fact is. For example, it is not clear if we should take this fact to involve possible worlds. However, my view is not committed to this factuality taking any particular form and so this is not something that I need to clarify.

A different problem for the idea that the j-reasons for emotions are facts stems from the idea that they are provided by entities that are more simple than facts. Specifically, the j-reason for an emotion might be an object. One might be led to believe that emotions can have reasons which are simple objects by considering love. It appears that one might simply love a person rather than loving any particular fact about them. However, love is not generally regarded as reasonable or unreasonable. When someone simply loves a person rather than anything about that person it is right to say that he or she does not love that person for a reason. He or she simply loves that person. Thus love is often singled out as quite unlike other emotions.[35] It is not a counter example to the idea that the j-reasons for emotions are facts. Rather, we normally cite facts as reasons for emotions.

It is true that emotions are often directed towards objects. For example, my anger might be directed towards my neighbour. I might act abruptly with him. But that does not mean that my neighbour is the reason for my anger. On the contrary, the reason for my anger must be something more complex—a fact. This is made clear by the fact that I can have a variety of emotions directed at my neighbour at the same time. In such a case I am not simply angry and grateful for the reason that my neighbour. (Note how odd it is to try to say one's reason is merely an object.) Rather, I am angry because my neighbour woke me up and I am grateful because he put out my bin. My reasons are facts that involve my neighbour.

9.5 Appropriateness and Tense

At this stage I would like to return to Prior's thank goodness argument that I discussed in Chap. 2. Prior suggested that when he said "Thank goodness that's over" he was expressing relief that the exams were over. He was not expressing relief that they finished on the 15th of June 1954 or that they finished simultaneously with his utterance. Therefore, the utterances "the exams are over" and "the exams end on the 15th of June 1954" or "the exams end simultaneously with this utterance" do not have the same meaning even if they are all uttered at the same time. Prior's relief that involved the tensed belief 'the exams are over' would be different from any relief he felt which involved the tenseless beliefs 'the exams end on the 15th of June 1954' or 'the exams end simultaneously with this relief'.[36] In this section I will push this distinction between what I shall refer to as the *tensed* and the *tenseless emotions of relief* a little further. I will also stress the distinction in regard to some other emotions.

Prior (1959, 17, 1962) not only denies that he has an emotion of relief that can be captured with tenseless language, he actually questions why anyone would have such an emotion. I take this to raise the issue not only that these tenseless emotions of relief would be odd, but further that they would be inappropriate.[37] Relief is orientated towards the past. It is appropriate to be relieved that a discomfort *has ended* but it is inappropriate to be relieved that a discomfort *will end*.[38] In a connected manner, it is inappropriate to have a tenseless emotion of relief because in such tenseless relief one will not know whether or not the discomfort has ended. Just as Prior's relief is inappropriate if the exams are not over, it is also inappropriate if he does not know whether they are.

One could have tenseless beliefs that a discomfort ends at a certain time or simultaneously with a certain feeling of relief/belief, and also not know whether that discomfort has ended. The first sort of case would arise if one knew that the exams ended at a specific time, but did not know what the current time was. An argument used in Chap. 3.4 can be used to make the other case clear. Suppose that mental states are brain states[39] which can be seen in brain scans. Further suppose that that most people have small portable brain scanners in their heads and that these states are all displayed in a lab. One might then point to a screen and refer to and have a belief/emotion about a pictured brain state without knowing it was one's own and if one did not know whether the broadcast was live, without knowing if the state was present. If this happened one could have a self-referential state about when a discomfort ends and lack the ability to make accurate tensed beliefs about it. One could know that the exams end simultaneously with a state one is looking at and not know if the exams have ended yet.[40] In this way emotions can be responsive to tenseless facts and ignorant of tensed ones. Tenseless relief thus lacks a sensitivity to whether a discomfort has ended yet it is precisely a sensitivity to this that makes a relief appropriate. (It is important to note that all parties must agree that it is true to say some things are past and others are not, and therefore these comments do not presuppose a tensed theory of time.)

Relief is only appropriate when orientated towards the past. Many other emotions are also temporally orientated. Some are orientated towards the past and others towards the future or present. Fear is only appropriate when one is or will be in danger.[41] This temporal orientation is embodied in the fact that one must be sensitive to time in having these emotions if one is going to be appropriate. One is only so sensitive if the emotions are tensed because tenseless emotions, precisely in being tenseless, are not so sensitive.

In the case of fear one can have the tenseless belief that 'I am in danger at *t*' and not know whether one is, will be, or was in danger if one doesn't know the time. One can also have the self-referential belief 'I am in danger simultaneously with this belief' and not know whether one is or was in danger. This is made clear by considering someone viewing her belief as brain state on a computer screen. The person may believe of the belief she sees that she is in danger simultaneously with it. But not knowing whether the image is live or not, she will not know whether the danger is present or past. As a result, she should not be afraid because it is not appropriate to be afraid of a past danger. (We should not be afraid because we were in danger last week.)

A consideration of the facts picked out by tenseless beliefs/emotions can also make it clear that they are not j-reasons for relief. If a discomfort of mine has ended I have reason to feel relieved and if it has not, I do not. I have reason to feel relieved quite independently of other factors. For example, it is irrelevant that the sky is blue, that I believe the sky is blue, or that I believe a myriad of other things. Similarly, if a discomfort of mine has not ended, I lack a reason to feel relieved quite independently of these other factors. More specifically, I have or lack a reason to be relieved whether or not I am relieved and therefore whether or not I have a state of relief after the ending of a discomfort. If a discomfort of mine has ended I have a reason to be relieved even if I am not relieved and I have a reason to be relieved even if I am relieved (relief doesn't make itself unreasonable). It thus appears that the fact picked by the tenseless self-referential belief is not a fact that makes my relief reasonable. That I am relieved after a discomfort is not a j-reason for me to be relieved.[42] Analogously, I have a reason to be relieved if a discomfort has ended and no reason to be relieved if a discomfort of mine has not ended. This is so quite independently of the date on which the discomfort ends. Therefore, the fact that the discomfort ends on a specific date is not a fact that makes it appropriate for me to be relieved.

Emotions can have a temporal orientation and hence can be such that they are only appropriate if they involve a tensed belief and not if they involve a tenseless one. Prior's relief is a case in point. Prior lacks a tenseless emotion of relief and he is quite right to lack one.

9.6 The Argument from Emotions

In this section I will bring together what has been revealed about emotions in order to provide an argument against the tenseless theory of time. This argument will bring out important aspects of Prior's thank goodness argument and build on arguments given in Chap. 3.

I have shown that whilst we can consider the practical value of an emotion this is not generally what is of concern when we consider the appropriateness of the emotion. We can also evaluate the belief involved in an emotion in terms of whether it is appropriately sensitive to the truth: for example, we can consider whether it originated from a source likely to lead to true beliefs. Again, however, this is not what is normally of concern when we consider the appropriateness of an emotion. If I am extremely angry because someone accidentally knocked my elbow we readily judge my emotion to be inappropriate. Further, we do this whilst unaware of whether the emotion is of any practical value and knowing that the cognitive content comes from a reliable source.

I showed that emotions were commonly evaluated as appropriate or not on the basis of the reasons that they were felt for. An emotion will be appropriate if it is felt for a reason that justifies an emotion of that type. The reason an emotion is had for is simply the fact which is captured by the belief that the emotion involves. This fits our everyday practice of giving reasons for our emotions or reasons which would justify being emotional.

I have also considered the pattern of appropriateness that arises in the case of some emotions. It was shown that there are a number of emotions that have a temporal orientation so that they can only be appropriate if they involve a tensed belief. This appropriateness concerns the reasons the emotions are felt for. It is determined irrespective of whether the belief involved is from a reliable source or whether the emotion has a practical value (the latter is often simply unknown).

Our consideration of emotions thus shows that when Prior is relieved because the exams are over he is relieved for a reason picked out by his belief 'the exams are over'. Further, this reason cannot be picked out by any of Prior's tenseless beliefs because the relief would not be reasonable if it involved a tenseless belief. Because the reasons at issue are facts, it follows that Prior's tensed belief picks out a fact not picked out by any tenseless belief.

A tensed belief will be a belief an agent has when that agent is disposed to honestly express or affirm tensed uses of language. It will be a belief that an agent communicates by using tensed language. It is therefore natural to take the beliefs and the language to share objects and to capture the same facts. Therefore, if tensed beliefs capture facts that tenseless ones do not, then tensed language captures facts that tenseless language does not. The definition of the tenseless theory is that it denies that tensed language captures facts that tenseless language does not. Therefore, the tensed theory is wrong.

Notes

1. By 'simple feeling' I mean a feeling devoid of cognitive content and therefore I am excluding from these comments theories such as Greenspan's (2004a), de Sousa's (1987), or Goldie's (2004) in which one might want to say that an emotion was a feeling with a cognitive content.
2. Cf. Solomon (1993, ch. 4).
3. Cf. Ryle (1949, ch. 4).
4. Cf. Solomon (1993, ch. 4) and more generally Davidson (1963/2001) and Lowe (2000, ch. 3).
5. Solomon (1993, 98). At this time Solomon was keen to defend the idea that an emotion is a judgement, however, later Solomon shifts to the view that emotions essentially involve judgments even if they also have other aspects, such as feelings (e.g. Solomon (2003)).
6. A number of writers have defended the idea that emotions have cognitive content, though not all of these writers are happy to speak in terms

of belief as providing that cognitive content. I will say something of this below. (Cf. Solomon (1993), Greenspan (1988), Calhoun (2003), and de Sousa (2004).) (This distinction assumes a narrow use of 'belief' not the umbrella one that I have been using.)

7. Cf. e.g. Solomon (1993, ch.5), Greenspan (1988), and Taylor (1975).
8. Cf. Pitcher (1965), who cites Bedford (1957) and Urmson (1952).
9. Cf. Shaffer (1983).
10. Solomon (1988) actually argues in response that the notion of pure belief or judgment is in fact unclear, insofar as there appears to be something at fault in or pathological about a belief that one has, e.g., been wronged, which lacks relations to or elements of feeling. I think this is interesting but it does not—as Solomon acknowledges—undermine the response I have given.
11. If one were to insist that an emotion was a feeling with a cognitive content, then I might be happy to speak of this feeling as a belief given an umbrella use of the term 'belief' as a mode of awareness (cf. Chap. 4). In this restricted sense an emotion may be no more than a belief.
12. Cf. Pitcher (1965), Deigh (2010), Greenspan (1988, ch.1), Calhoun (2003), and de Sousa (2004).
13. Cf. Deigh (2010).
14. Cf. Greenspan (2004b), Solomon (2003), and de Sousa (2004).
15. Deigh himself supposes something like this (2010, 31).
16. Cf. Solomon (1998).
17. It should be noted that Robinson does not take herself to have proven that no emotions involve a cognitive element, merely that not all do.
18. Though that is not to say the idea is new. Aristotle was perhaps an earlier example (cf. Aristotle (384-22BC/2001, bk. IV)).
19. Cf. Greenspan (2004a). She refers to the two forms as emotions *adaptiveness* and *appropriateness* respectively.
20. E.g. Solomon (1993, ch. 4) and Pitcher (1965).
21. A different social practical advantage of emotions might be taken to arise from the idea that expressions associated with emotions aid social cohesion (cf. Robinson (1995) who attributes such an idea to Paul Ekman). But, given that emotions and behaviour needn't be covariant, then this appears to be implausible.
22. Cf. Nagel (1978, 71 ftnt. 1), Maclaurin and Dyke (2002), and Dyke and Maclaurin (2013).

23. It is no good to suppose that an emotion must be of practical value otherwise it would have been ruled out by evolution, because it could be that emotions are side effects of things which are of value, that emotions are of neutral practical value, or that we have not evolved to a state at which the negative practical value of emotions has knocked them out of existence.
24. It is of note that Maclaurin and Dyke (2002) admit that they struggle to explain grief in terms of practical value despite trying to account for all emotions in this way.
25. de Sousa (1987) speaks of the appropriateness of an emotion in terms of whether the context of the emotion is relevantly similar to the paradigm context of the emotion. But, this does not offer us an alternative to a practical assessment of emotions because de Sousa takes the appropriateness of the paradigm contexts to be determined by practical matters.
26. Cf. Bergmann (1978).
27. E.g. Audi (2004). My argument is not reliant on this view of the appropriateness of beliefs.
28. Greenspan, e.g. (1988) says that the appropriateness of an emotion will not be identical with the appropriateness of the belief it involves. Her reason for saying this is that the belief will often have to meet higher standards to be appropriate than the emotion does. However, this difference in level changes little of relevance to my argument.
29. I am not the only person to speak of reasons in regard of the appropriateness of emotions (cf. Pitcher (1965), Taylor (1975), Goldie (2004), and Solomon (1993)). However, I think that it is important to carefully separate this from considering the practical value of an emotion and considering whether the belief involved in the emotion is appropriate.
30. Solomon believes that the emotion will be triggered by something objective, but that objective thing needn't be anything to do with the fact that my headache is over.
31. E.g. Greenspan (1988), Calhoun (2003), de Sousa (2004), and Pitcher (1965) (though Pitcher does not make it a general claim).
32. Note, the more one takes there to be evidence that the emotions at issue are had for a reason the more evident it becomes that there is irrationality involved. If this evidence should turn out to be compelling, then we ought to bite the bullet and accept the irrationality. If so, we still lack grounds for weakening the cognitive component of an emotion.

33. Some have tried to criticize the cognitive view on the grounds that some of the physiological aspects of an emotion start independently of cognition, because they start before cognition starts. In short, there are physiological reactions to environmental stimuli before these stimuli are cognised (cf. Damasio (2000)). This criticism is problematic because it is not clear what is at issue in the notion of cognition (cf. Solomon (2004)). Moreover, this does nothing to undermine my argument even if it turns out these initial physiological reactions are identical in the case of normal and recalcitrant emotions. This is because one can distinguish these initial physiological conditions from emotions proper, which must be something more. Arguments in support of the cognitive theory can therefore be seen as arguments in support of the idea that emotions are more than these initial physiological conditions. Normal and recalcitrant emotions might therefore still differ in their responsiveness to reasons, even if their initial physiological conditions do not differ in their reaction to stimuli.
34. Cf. Brady (2009, 428) and Audi (2004) who was discussed above.
35. E.g. Pitcher (1965).
36. I am assuming the demonstrative 'this' is tenseless though denying this would do little to alter my argument.
37. Cf. Cockburn (1998, 83).
38. Even if it is appropriate to be relieved to have learnt that a discomfort will end, e.g., be relieved when one's discomfort at the prospect of endless discomfort ends. Hoerl (2013) in fact distinguishes two forms of relief: temporal and counterfactual. If this distinction were accepted my argument might only apply to the former sort but this is the sort had by Prior and therefore it is the sort that is relevant here.
39. My argument does not rest on this assumption. It merely rests on the idea that mental states can be known third-personally as well as first-personally.
40. Even if one could narrow the options down to being such that the state being looked at was either present or past this will still not be sufficient to know the discomfort has ended. Further, the pattern of temporal orientation manifested in relief is quite different to that manifested in other emotions and so this cannot be a general response to the matter at hand, as I will make clear shortly.
41. Cf. Cockburn (1998) and Maclaurin and Dyke (2002).

42. Well, it might be a reason in a bizarre case in which I am concerned about whether or not I can feel relieved. However, the everyday case such as Prior's is not like this.

References

Anscombe, G. E. M. (1957/2000). *Intention*. London: Harvard University Press.

Aristotle. (384-22BC/2001). Nicomachean Ethics (Ethica Nicomachea) (W. D. Ross, Trans.). In R. McKeon (Ed.), *The Basic Works of Aristotle* (pp. 935–1112). New York: The Modern Library (Random House).

Audi, R. (2004). Theoretical Rationality: Its Sources, Structure, and Scope. In A. Mele & P. Rawling (Eds.), *The Oxford Handbook of Rationality* (pp. 17–44). Oxford: Oxford University Press.

Bedford, E. (1957). Emotions. *Proceedings of the Aristotelian Society, 57*, 281–304.

Bergmann, F. (1978). Review of 'The Passions'. *Journal of Philosophy, 75*, 200–208.

Brady, M. S. (2009). The Irrationality of Recalcitrant Emotions. *Philosophical Studies, 145*, 413–430.

Calhoun, C. (2003). Cognitive Emotions. In R. C. Solomon (Ed.), *What Is an Emotion? Classic and Contemporary Readings*. Oxford: Oxford University Press.

Cannon, W. B. (1917). The James-Lang Theory of Emotions: A Critical Examination and an Alternative Theory. *American Journal of Psychology, 39*, 115–124.

Cockburn, D. (1997). *Other Times*. Cambridge: Cambridge University Press.

Cockburn, D. (1998). Tense and Emotion. In R. Le Poidevin (Ed.), *Questions of Time and Tense* (pp. 77–91). Oxford: Oxford University Press.

Damasio, A. (1995). *Descartes' Error*. London: Picador.

Damasio, A. (2000). *The Feeling of what Happens: Body, Emotion and the Making of Consciousness*. London: Vintage.

Davidson, D. (1963/2001). Actions, Reasons, and Causes. In *Essays on Actions and Events* (pp. 3–19). Oxford: Oxford University Press.

de Sousa, R. (1987). *The Rationality of Emotion*. Cambridge, MA: Massachusetts Institute of Technology Press.

de Sousa, R. (2004). What I Know, What I'd Like to Think I Know, and What I'd Like to Think. In R. Solomon (Ed.), *Thinking about Feeling: Contemporary Philosophers on Emotions* (pp. 61–75). Oxford: Oxford University Press.
Deigh, J. (2010). Concepts of Emotions in Modern Philosophy and Psychology. In P. Goldie (Ed.), *The Oxford Handbook of Philosophy of Emotion* (pp. 17–40). Oxford: Oxford University Press.
Dyke, H., & Maclaurin, J. (2013). Evolutionary Explanations of Temporal Experience. In H. Dyke & A. Bardon (Eds.), *A Companion to the Philosophy of Time* (pp. 521–534). Oxford: Wiley Blackwell.
Elster, J. (2004). Emotion and Action. In R. Solomon (Ed.), *Thinking about Feeling: Contemporary Philosophers on Emotions* (pp. 151–162). Oxford: Oxford University Press.
Evans, D. (2002). The Search Hypothesis of Emotions. *British Journal for the Philosophy of Science, 53*, 497–509.
Goldie, P. (2004). Emotion, Feeling and Knowledge of the World. In R. Solomon (Ed.), *Thinking about Feeling: Contemporary Philosophers on Emotions* (pp. 91–106). Oxford: Oxford University Press.
Greenspan, P. S. (1988). *Emotions and Reasons*. London: Routledge.
Greenspan, P. S. (2004a). Practical Reasoning and Emotion. In A. Mele & P. Rawling (Eds.), *The Oxford Handbook of Rationality* (pp. 206–221). Oxford: Oxford University Press.
Greenspan, P. S. (2004b). Emotions, Rationality, and Mind/Body. In R. Solomon (Ed.), *Thinking about Feeling: Contemporary Philosophers on Emotions* (pp. 125–134). Oxford: Oxford University Press.
Hoerl, C. (2013). Tense and the Psychology of Relief. *Topoi* (special issue on *Time and Time Experience*, eds. G. Torrengo & R. Ciuni): 1–15.
Hume, D. (1739–40/1985). *A Treatise of Human Nature*. London: Penguin.
James, W. (1884). What is an Emotion? *Mind, 9*, 188–205.
Lowe, E. J. (2000). *An Introduction to the Philosophy of Mind*. Cambridge: Cambridge University Press.
Maclaurin, J., & Dyke, H. (2002). Thank Goodness That's Over: The Evolutionary Story. *Ratio, 15*, 276–292.
Nagel, T. (1978). *The Possibility of Altruism*. Chichester: Princeton University Press.
Pitcher, G. (1965). Emotion. *Mind, 74*, 326–346.
Prior, A. N. (1959). Thank Goodness That's Over. *Philosophy, 34*, 12–17.
Prior, A. N. (1962). The Formalities of Omniscience. *Philosophy, 37*, 114–129.
Robinson, J. (1995). Startle. *Journal of Philosophy, 92*, 53–74.

Ryle, G. (1949). *The Concept of Mind*. London: Hutchinson.

Shaffer, J. A. (1983). An Assessment of Emotions. *American Philosophical Quarterly, 20*, 161–174.

Solomon, R. C. (1988). On Emotions as Judgments. *American Philosophical Quarterly, 25*, 183–191.

Solomon, R. C. (1993). *The Passions: Emotions and the Meaning of Life*. Cambridge: Hackett.

Solomon, R. C. (1998). The Politics of Emotion. *Midwest Studies in Philosophy, 22*, 1–20.

Solomon, R. C. (2003). On the Passivity of the Passions. Reproduced in *Not Passions Slave* (pp. 195–232). Oxford: Oxford University Press.

Solomon, R. C. (2004). Emotions, Thoughts, and Feelings: Emotions as Engagements with the World. In R. Solomon (Ed.), *Thinking about Feeling* (pp. 76–88). Oxford: Oxford University Press.

Taylor, G. (1975). Justifying the Emotions. *Mind, 84*, 390–402.

Urmson, J. O. (1952). Parenthetical Verbs. *Mind, 61*, 480–496.

10

Time and Self

In the preceding chapters I have provided two arguments to show that the tensed theory of time is correct. I am not the first to argue for this conclusion and McTaggart's paradox has provided a popular response to it. In this chapter I will outline McTaggart's paradox and show that it does not refute the tensed theory of time.

My arguments for the tensed theory of time made it clear that tensed beliefs/language pick out facts not picked out by tenseless beliefs/language. In Chap. 7 I made it clear that these arguments can also be used to show that first-personal beliefs/language pick out facts not picked out by non-first-personal beliefs/language. I will here consider how this conclusion could support emergentism about the self. I will also outline Wittgenstein's private language argument which might be thought to raise problems for this conclusion. I will not provide a full solution to Wittgenstein's argument but I will argue that any problems that it raises for my conclusion are not decisive.

Olley (F.O.C.H.) Pearson, *Rationality, Time, and Self*,
https://doi.org/10.1007/978-3-319-71973-3_10

10.1 The Tensed Theory of Time

As I noted in Chap. 2, the tensed theory of time is not primarily a theory about language. It is also not a theory about grammar or a theory that takes time to have linguistic characteristics. The tensed theory, as I have characterized it, is that tensed language picks out facts not picked out by tenseless language. This is a general characterisation.

The tensed theory can take more specific forms. For example, one might suppose that 'now' picks out a property of presentness. Tensed language, unlike tenseless language, would inform us of the instantiation of this property. Alternatively, one might think that 'now' picks out a limit of existence so that tensed language tells us something about what exists that tenseless language does not. I am deliberately not making more specific claims about the nature of time. I have argued a tensed theory is correct but not which tensed theory is.

Various arguments have been given to support tensed theories. One prominent example stems from experience. Our experience is of a world in which what is present is continually changing. It appears hard to account for this experience if the tenseless theory is correct because tensed descriptions appear to be unchanging in truth-value. To this extent my argument is in good company. However, various arguments have also been given against tensed theories. In the following section I will discuss what I take to be the most prominent of these and I will show that it is inconclusive.

10.2 McTaggart's Paradox

McTaggart's paradox, as I shall refer to it, is the most discussed criticism of the tensed theory of time. In this section I will present it and show that there are a number of ways the tensed theory can respond to it.

As we saw in Chap. 2, McTaggart spoke of the A- and B-series, and A- and B-characteristics. The A-series is the series running from the distant past to the far future and A-characteristics are the characteristics of presentness, pastness, and futurity. The B-series is the series from earlier to later and the B-characteristics are earlier/before, simultaneous with, and later/after. The latter captures tenseless time and the former tensed time.

McTaggart (1908, 1927, ch. 33) argued as follows:

P1, time requires change;
P2, change requires the A-series;
P3, the A-series is impossible;
C1, therefore time is impossible.

McTaggart provides little argument in support of P1. Whilst there is scope to question it[1] I will not do so here because I believe the heart of the matter lies elsewhere.

McTaggart took the constituents of times to be events and his defence of P2 assumes that if there is change, it must be the changing of events. He argues that events cannot change by coming into or out of existence as their order in the B-series is unchanging. Events also cannot change most of their properties or constituents. For example, the sun rising at 4am must occur at 4am and must be a rising of the sun otherwise it is a different event. The only characteristics of an event that can change without changing the identity of the event are its A-characteristics. A sunrise can pass from the future into the past and remain the same event. Thus McTaggart concludes change requires the A-series.

McTaggart notes that the A-characteristics are incompatible. He adopts the following:

A1. If an event is past, it is not also present or future. If it is present, it is not future or past. If it is future, it is not present or past.

(Things can of course be extended in time, but if we make more fine grained distinctions McTaggart's point seems right. For example, if the beginning of the war is past, it is not present or future. I'll therefore ignore temporal extension as a counter example in my discussion.) However, McTaggart believes that every event must have all three characteristics.

A2. Every event is past, present, and future.

Events must pass from being future to being present and then past. (If there is a beginning or end of time some events might have only two characteristics but this changes little). Thus McTaggart believes that the A-series is self-contradictory and hence impossible, P3.

In considering how the tensed theory can respond to McTaggart's paradox it is fruitful to bear in mind that it is often supposed that the tenseless theory can avoid the paradox but the tensed theory cannot (and therefore the tenseless theory is right).[2] The tenseless response to the paradox denies P2 and hence C1. It concludes the B-series can exist but the A-series cannot.

Russell (1903/1996, sect. 442) provides a very different account of change to McTaggart's. For Russell it is things rather than events that change. For example, a tomato changes from green to red as it ripens. Specifically, a thing changes if it has a property at one time that it does not have at another time.[3] This account of change appears not to require the A-series and so can avoid McTaggart's argument by denying P2.

McTaggart considers such a response to his argument. He criticizes this account of change on two grounds. Firstly, it doesn't appear to provide change because the facts remain unchanging. Secondly, it does not account for why change does not occur in space in an analogous way. Why do we not have change when an object has a property at one place that it does not have at another place? (Imagine a poker partially in a fire with one hot end and one cold end).

However, McTaggart's rebuttal against a Russellian theory of change is weak. His first point merely presupposes the account of change under consideration is wrong and thus has no force. McTaggart simply asserts that there is no change when a Russellian would take there to be change and a Russellian does not need to be moved by that. His second point, whilst interesting, can be met because there are asymmetries between time and space. For example, causation happens across time but not necessarily across space.[4] Therefore, it is open to someone who adopts a Russellian theory of change to say that time is fundamentally different from space and this is why we have change when an entity has different properties at different times but not when it has different properties at different places.

The tenseless theory could thus adopt a Russellian account of change and thereby deny P2. Nevertheless, it is also open to a tensed theory to adopt such a theory of change but to insist upon P2. If one has independent reasons for thinking time is tensed then these are reasons for thinking that the references to time made in the Russellian definition of change must be references to a tensed time or an A-series.[5]

McTaggart's defence of P3 assumes A1 and A2. However, A2 is clearly false.[6] Nothing is past, present, and future. My sitting down to supper this evening is future but not present or past. My writing this is present but not future or past. My birth is past but not future or present. Therefore, P3 ought to be denied. Moreover, this should be clear from the plausibility of the tenseless account of the truth-conditions of tensed sentences. According to this account a token of '*e* is present' said at *t* is true if and only if *e* occurs at *t*, a token of '*e* is past' said at *t* is true if and only if *e* is before *t*, and a token of '*e* is future' said at *t* is true if and only if *e* is after *t*.[7] Therefore on the tenseless theory a token of '*e* is past, present and future' said at *t* is true if and only if *e* is before, simultaneous with, and after *t*. Because we are neglecting the temporal extension of *e* to make A1 plausible these truth-conditions can never be fulfilled. So, A2 is clearly false.

Denying A2 and hence P3 undermines McTaggart's argument and prevents the paradox being a problem for the tensed theory of time. Nevertheless, I will say a little more in order to explain away the temptation to accept A2.

McTaggart supports A2 with reference to his theory of change according to which events pass from future to past. However, we have already seen that we needn't accept McTaggart's theory of change. Furthermore, this theory of change does not entail an event is past, present, and future. At best it encourages one to say the following:

> A6. Every event is past and was present and was earlier still future, or will be past, is present, and was future, or is future and will be present and subsequently past.

Any of the disjuncts of A6, and hence A6 itself, is compatible with A1.

Nonetheless, McTaggart does consider A6 and any of its disjuncts to be incompatible with A1. McTaggart considers his opponent might try to deny A2 and replace it instead with a disjunct of A6. "My birth is not past, present, and future, rather it is past, was present, and was before that future." McTaggart argues that this move from A2 to a disjunct of A6 is inadequate to remove conflict with A1 because it entails either a vicious circle or a vicious infinite regress. McTaggart says this because he takes saying "*e* will be past" to be equivalent to saying "*e* is past in the future" and so on for the other tenses. So McTaggart believes that moving from A2 to A6 or one of its disjuncts is simply moving from attributing one set of A-characteristics to attributing two sets. But, given attributing one set was a problem attributing two sets is also a problem.

However, McTaggart is wrong in this. We don't need to say "*e* is past, present and future" to say "*e* is past", "*e* is present", or "*e* is future". We don't need to adopt A2 to adopt the tensed theory or A-series. Attributing one set of A-characteristics is not a problem. Similarly, saying A6 or one of its disjuncts does not require us to say the following:

A7. *e* was present, is present, and will be present (or: *e* is present in the past, present, and future).

A7 could be problematic and could be a step on a vicious circle or infinite regress, but A6 and its disjuncts are not.

Because attributing one set of A-characteristics is not problematic attributing two sets of A-characteristics does not inherit any problems from it. However, it might seem that attributing a second set of A-characteristics in this way is problematic because it requires there to be a second time series. Further, one might worry that if we have a second time series we will arrive at the same issues with it as we did with the first and so will be led to attribute a third time series and so on to infinity. But, this is not the problem it might seem. Firstly, there is no problem with having a time series so this would not be a vicious regress. At most this is not very ontologically parsimonious.[8] Secondly, there is no need to venture on a regress. As Prior (1967, 19) says, the 'was' and 'future' in 'e was future' appear to be just our usual ones. So, we do not need a second time series for A6 or one of its disjuncts to be true.

Lowe[9] denies that one even ought to say A6 or something equivalent on the grounds that one cannot iterate indexicals. It makes no more sense to say "*e* will be past" or "e is past in the future" or "*e* is after now before now" than it does to say "Lampeter is here there" or "I am John at you".[10] (This links with my comments concerning the truth-conditions of tensed utterances.) Lowe explains away any temptation to say something like A2 or A6 by providing an alternative which avoids the pitfalls just alluded to. There is a true and meaningful metalinguistic alternative: if it is true to say "*e* will happen", it will be true to say "*e* is happening" and subsequently "*e* has happened". This metalinguistic alternative captures everything that erroneously appears right in A2 or A6.

McTaggart's paradox is no problem for the tensed theory as A2 and hence P3 are false. It is not true that every event is past, present, and future, and therefore it is not true that the A-series is impossible. Further, one should not be tempted to A2 or P3 by McTaggart's theory of change.

10.3 The Emergent Self

The argument from action establishes that first-personal beliefs and language can capture facts not captured by non-first-personal beliefs or language. This relates to dualist and emergentist theories of the self.

The physical body appears to be fully describable by a future empirical science. Science aims at objective observer independent descriptions of the world. It is therefore natural to suppose that the physical body could be fully described with non-first-personal language. Physicalism is a view according to which there is nothing more to the self than the physical body. This implies that according to physicalism the self can be fully described with non-first-personal language.

If first-personal beliefs/language capture something which is not captured by non-first-personal beliefs/language, then plausibly this unique element concerns the self. If there is something unique about me or I, it is something unique about my self.

To this extent the argument from action conflicts with physicalism and supports a form of dualism. There is something more to the self than the physical body. For example, the self is a distinct substance from the physical body or it has characteristics the physical body lacks.

Various arguments have been given in support of forms of dualism. For example, Lowe (2006b, 2008) has argued that I am the unique subject of all of my thoughts and feelings. However, this is not true of my body as a whole or of any part of it and therefore I am not my body or any part of it. For example, I must exist to believe 'I am happy'. But I can believe this without my whole body existing so I am not my whole body. It might be necessary that part of my brain must exist for me to have that belief. But, that same part is plausibly not also necessary for all of my other mental states. I might feel pain even if that part of my brain did not exist. Therefore, I am not identical with that part of my brain. Similar remarks apply to my other mental states.

Dualism is the case if the self is something over and above the physical body or if there are characteristics of the self which are over and above the characteristics of the body. It requires entities which are distinct from the physical body and/or its characteristics. Emergentism also involves this. An emergent entity is distinct from an entity it emerges from. However, emergentism also has a second aspect. An emergent entity will depend in some way on the entity it emerges from.[11] This notion of dependence can be understood in different ways and I will not specify one here.[12] However, it is plausible that dependence of some form occurs in the current case. Facts picked out by first-personal beliefs/language plausibly depend on facts picked out by non-first-personal beliefs/language. At the least, there appears to be some form of covariance between the two. The truth of first-personal beliefs/utterances do not appear to be able to vary entirely independently of the truth of non-first-personal beliefs/utterances. To this extent the argument from action appears to support a form of emergentism about the self.

Nevertheless, it is of note that the argument from action has implications that are not shared by all dualist or emergentist views. The facts picked out by my first-personal beliefs/utterances are plausibly distinct from those picked out by the first-personal beliefs/utterances of other people. I might truly believe 'I am making a mess' whilst no one else can truly believe that. In this way the facts captured by my first-personal beliefs involve an element of privacy. They are not captured by anyone else's first-personal or non-first-personal beliefs. This privacy is not an essential commitment of dualism or emergentism. I will consider the plausibility of this privacy in the following section.

10.4 The Private Language Argument

The first-personal fact that I am aware of when I correctly believe 'I am making a mess' is a fact that cannot be captured by any non-first-personal beliefs or by anyone else's first-personal beliefs. To this extent the fact is knowable only to me. The implication of this is that when I utter "I am making a mess" I express a fact that no one else can take up as the object of a belief in this way. To this extent I appear to be saying something private. Wittgenstein (1953/2001) has argued[13] against the possibility of a private language and in this section I will consider this argument and how it bears on my position. (I will refer to Wittgenstein's argument as the *private language argument* or *PLA*.)

It is commonly thought that the core of the PLA appears between sections 243 and 271 of the first part of Wittgenstein's *Philosophical Investigations*.[14] In Section 243 Wittgenstein makes it clear what he means by private: the words refer to things that can only be known to the speaker such as his or her sensations.

The focus of the argument appears in section 258 where Wittgenstein invites us to imagine a scenario in which he is trying to record the reoccurrence of a particular sensation of his. He writes the letter 'S' in his diary on the days he has the sensation. Wittgenstein says that the sign—'S'—cannot be defined. Instead he supposes that he attempts to establish its meaning using an ostensive definition. He cannot point to the sensation in the way we normally point at things external to us. Instead Wittgenstein supposes that he performs a form of inner pointing by focusing his attention on the sensation when he speaks or writes the sign. However, at this point Wittgenstein turns to considering this scenario and the value of this inward pointing. The pointing establishes the connection between the sign and the sensation so that Wittgenstein will remember it correctly in the future. But, Wittgenstein suggests that there is no criteria of correctness in this case. On the contrary, whatever seems right to him will be right. This leads him to say that we cannot really talk about the right use of the sign in this case.

There have been a great number of different interpretations given of the PLA and I will begin by expanding on these.[15] The idea that the diarist must remember the connection right can be variously understood. It might be taken as a reference to the idea that we appear to have a certain incorrigibility as regards our own sensations: I can be wrong about the cause of my pain but not that I have pain. This gives us the following:

> U1. If the diarist is to establish meaning, then he must establish something involving incorrigibility.[16]

Alternatively, one might consider that the act at issue is to establish the meaning of the sign for the individual giving us the following:

> U2. If it is possible for the act to occur and the individual to fail to grasp the meaning, then the act did not succeed and therefore is not after all an act which establishes meaning.[17]

The idea that in the present case there can be no criteria of correctness can also be variously interpreted:

> U3. One might think that there is no criterion for checking that the sensation had when 'S' is first used is the same as that had when it is later used.
> U4. Or, one might think that there is no criterion to verify that the meaning 'S' had when first used is the same as it is taken to have when later used.[18]
> U5. Or, one might doubt whether there was a meaning established in the first place and hence any sort of criterion of correctness.

There are no indicators of the presence of S other than the diarist's awareness of S. Therefore, when the diarist tries to confirm that a sensation he has now is the same as that which he had when he first used 'S' all he can compare his current sensation with is the memory of the past sensation. However, this memory can only act as a criterion of correctness if it can itself be verified. There is nothing against which the memory can be checked and therefore the diarist has no criteria for recognizing that his later sample is the same as S, U3. Similarly, the diarist will have no access to the original meaning of 'S' other than through his memory which cannot itself be verified by the diarist and hence cannot provide a criterion for checking that a later use of 'S' shares meaning with the original use, U4.

U3 and U4 thus imply that the diarist can be mistaken in his later use of 'S' because memory is not infallible. Thus they conflict with U1 and/or U2. The idea of the diarist and hence of a private language appears to contain an inconsistency and so to be impossible.

U3 and U4 might also be taken to be worrying in their own right on the following grounds:

U6. One might believe that an expression can only be meaningful if it is possible to verify whether or not it is used correctly.
U7. Or, one might believe that one can only learn to use or continue to use an expression meaningfully if he or she can verify when that expression is used correctly.[19]
U8. Or, one might think that the memory of the first experience of S was supposed to provide the diarist with the meaning in the later use but it can only do that if he knows which sensation to remember and that presupposes already knowing what the meaning of 'S' is.[20]

If any of these theories are right 'S' lacks meaning and a private language thus looks impossible.

One might believe that a criterion of verification alluded to in the previous paragraph was a general requirement on all uses of language because all uses of language must accord with a rule and the existence of a rule requires the possibility of verification.[21] Alternatively, one might think that although this isn't a general requirement the diarist sets it up as a requirement of his position because he sets up a case in which 'S' might be misapplied.[22]

U5 perhaps becomes most plausible when one also bears in mind other aspects of Wittgenstein's position. Focusing one's attention on a sensation and a sign might simply be inadequate to provide meaning. This could be because the possibility of naming something presupposes a great deal of ground work such as a practice of naming. Alternatively, it might be that one thinks that introspecting a sensation and ostensive definition generally are not alone sufficient to establish the identity criteria of a referent. An ostensive definition alone cannot delimit a pattern of correct use or a practice for a sign but these are essential for it to be meaningful.[23]

Thus it can be seen that there are a number of arguments against a private language that one might think of under the heading of the 'PLA'. At this stage I wish to raise an initial response. If the PLA is right, then we ought to conclude that Perry's utterance "I am making a mess" does not express a fact known only to Perry. However, this is clearly a meaningful utterance. Moreover, as noted above, it is clear that Perry's making of this utterance informs us that he has a belief different from one which he had before he was willing to make that utterance and different from any non-first-personal belief of his. We can inquire as to the nature of this belief and we can apply the argument given in Chap. 7 to conclude Perry must be aware of a fact not known non-first-personally. Crucially, however, this does not leave us with a contradiction or impossibility. It merely implies that not all public facts can be expressed non-first-personally. It would also follow that when I said "the fact Perry is aware of when he uttered "I am making a mess"" I was not referring to a private fact. But this needn't undermine my point. My argument would establish that there is something more than the facts we can know non-first-personally but it would not tell us what that thing or those things are. The ability to gesture beyond what is known is different from the ability to refer to something 'unknown'. At least some interpretations of the PLA will allow the former even if not the latter. There doesn't appear to be an analogue to U1 or U2 in the former case and nor is ostensive definition required. My argument and the PLA do not contradict and could both be right.

There are, however, good grounds for finding the PLA unconvincing. U1 perhaps takes incorrigibility to be required as it takes our everyday talk of sensations as a paradigm. But it is not clear that all sensations must involve incorrigibility and very doubtful that all private objects must. It was a common assumption that we have epistemic privilege, including incorrigibility, regarding our mental states.[24] But, this idea has met with extensive criticism.[25] For example, a close friend might make me recognize that I am angry without my previously being aware of that fact. Alternatively, my psychiatrist might force me to recognize that I am jealous although I was previously convinced that I was not. There is no reason to suppose that our knowledge of private objects is incorrigible. Private evidence might be reconsidered and accumulate, and hence lead to changes in judgement. Further, judgements based on private evidence can be susceptible to other influences: we can lie to ourselves.

U2 is implausibly restrictive. An ostensive definition could establish meaning even if people are sometimes forgetful. One's knowledge of a meaning needn't be unfaltering for one to know that meaning or for the meaning to have been established. On the contrary, I might introduce a secret language amongst a group of close friends even if only some of them grasp the meanings on first pass.

U6 supposes a form of verificationism, the idea that something must be verifiable to be meaningful. But this is an idea that ought to be abandoned. After all, it is an idea that cannot itself be verified and so it is self-refuting. It is similarly implausible that it must be possible to verify that someone is continuing to use a sign with the same meaning for that person to do so. It is possible for someone to act in a way that accords with a rule without anyone being able to verify as much. There can be patterns that cannot be verified. (Our epistemic restrictions do not determine the nature of the world. The world is unchanged by our knowledge or lack of knowledge of it.)[26]

It is also not required that the use of a sign can be verified in order for someone to learn the use of that sign. This might be so if all learning involved a process of trial and error, but it doesn't seem that it must. Language acquisition does not require us to repeatedly use words inappropriately until we are warned off all erroneous uses. It is not a blind fumbling. Further, one can adopt a practice without learning that practice. We can behave in ordered ways without having learnt those orders. Moreover, the diarist is not learning but creating meaning in this instance. Therefore, U7 is implausible.

Law (2004) pushes the case against U6, U7, and other similar principles by considering an example. We are invited to consider a man alone in a prison cell. This man sees rats occasionally and decides to record how often. He does so by writing a '!' in his diary on each day he sees a rat.

Now, after several months the prisoner might suddenly have a moment of doubt and wonder if '!' was marking days with or without rat sightings. The prisoner may have no way of getting a definite answer to this. Nevertheless, this lack of ability to determine the rule for the use of '!' clearly does nothing to undermine the fact that up until that point the prisoner was acting in a way that accorded with a rule. His behaviour followed a specific pattern even though he may not be able to verify what that pattern was. As Law stresses, the prisoner appears to need some sort of skill or ability to use '!' in a consistent way. However, he does not require opportunities to verify patterns of use. The case of the diarist is different to that of the prisoner. No one can join the diarist as the prisoner could be joined. But, this is of little relevance. The diarist could possess the relevant skill or ability to act in a consistent way just as the prisoner could.

There is no necessity for the diarist or the private language theorist to enter into the circle highlighted in U8. One does not need to be able to call to mind a sample of the referent of a sign before one can know what the sign means. If one had to do this it would cause trouble for all referring terms. It is also not the case that the diarist is committed to this being peculiarly how things are in the case of 'S'. Recalling a sample of the sensation is not supposed to be what fixes the meaning of the sign 'S'. On the contrary, as U8 implies, being able to recall a sample plausibly presupposes knowing the meaning.

U3 and U4 are also implausible. Memory can itself act as a source of verification. Memory is generally reliable and therefore remembering something to be the case can act to verify that it is the case. Further, there could be other sources of verification available. Blackburn[27] points out that if the diarist in fact kept note of a number of his sensations, then his diary may indicate a pattern. For example, S always occurs with T and after R. Such a regularity could provide the diarist with grounds for thinking he has mischaracterized a sensation on an occasion. For example, if he notes an occurrence in his diary of 'R' followed by 'T' and 'Q', then he will have grounds for thinking that the 'Q' was mistakenly applied to an S. The regularity will itself have been empirically confirmed. But, if it is well established and embedded in a theory of these sensations there could be grounds for holding onto that regularity and rejecting outliers. Moreover, the fact that the sensation is private does not entail that the sensation cannot leave traces which the diarist can be aware of and which can provide a criterion of verification. Traces can provide evidence. Ashes are evidence that there was a fire.

When considering U5 it is important to bear in mind that Wittgenstein believed that a person could follow rules and have a language even if he or she was and always had been alone. I will refer to such an individual as a *Crusoe* character. Section 243 from which I took Wittgenstein's notion of private actually starts by discussing individuals acting alone. There Wittgenstein says that a person can talk to, give orders to, ask questions of, answer, encourage, blame, etcetera, himself or herself.[28] Moreover, Wittgenstein seems right in this. We can imagine Crusoe, somehow alone on an island from birth, following practices and using signs meaningfully. Crusoe might scratch a cross onto a rock near streams that he has drunk from and been made ill by, and a circle next to those which have not made him ill. Hence, either ostensive definition alone can lead to meaningful language or there is a source of meaning besides ostensive definition and linguistic definition, and which does not require the cooperation of other people. Either way, U5 is no longer plausible. (Hanfling (1984) extends this point by making use of Wittgenstein's own notion of a family-resemblance. He shows that although 'S' is unlike 'pain' one can trace family-resemblances from 'pain' to 'S'. For example, steps can be made from 'pain' to 'S' through descriptions of dreams and phrases for sensations such as 'pins-and-needles'. Thus, both 'S' and 'pain' are meaningful.)

This brings to the fore Wittgenstein's notion of privacy. Privacy is not simply a matter of only being known to one person. It is more specifically a matter of having no outward appearance apart from the private linguist's awareness of it. Thus Wittgenstein speaks of what can only be known to the diarist and invites us to imagine sensations with no outward signs.[29] Privacy in this strong sense is thus crucial to the PLA because it is supposed to be what the diarist has but Crusoe lacks. I am hence encouraged to offer the following different reading of the PLA which I will refer to as *PLAI*: *For a sign to be meaningful it must have a use. It must play a role in a practice.*[30] *For a sign to refer to an object that object must play a role in this practice. It must influence the use of the sign.*[31] *Therefore, if a sign referred to an object the use of that sign would indicate the existence of that object. Any object referred to therefore is not private because it would appear in something external to it and one's awareness of it. Conversely, if an object was private, it could not influence the use of a sign and the use of the sign would be invariant to it. In such a case there is little ground to say that the sign refers to it.*

This does render Wittgenstein's argument very simplistic and I am not insisting this is exactly what Wittgenstein meant. But, PLAI would still have a bearing on those who are often recognized as Wittgenstein's targets, namely those who might think that for all anyone else can know I might have no sensation at all when I see red. Such a person would have to deny my red sensation had any influence on the use of the word 'red'.

PLAI does not succumb to the criticisms I raised above against the PLA. However, PLAI also does not conflict with the existence of first-personal facts because these facts are not private in the strict sense. Rather, these facts provide reasons for action and belief, and are thereby manifest in human practices. To this extent it is appropriate to take Perry's "I am making a mess" to express a fact not expressed by any other non-first-personal utterance or anyone else's first-personal utterances. A first-personal fact[32] is unlike S if S is supposed to be strictly private.

But, some of the things Wittgenstein says might also make it appear that there is a difference between a first-personal fact and the sensation of pain. Wittgenstein at times denies that there is an inner sensation of pain to the extent that "I am in pain" should not be understood as expressing a relation between two objects (myself and pain). The idea is captured in section 271. There Wittgenstein invites us to imagine a person who cannot remember what sensation the word 'pain' refers to and so he often picks out different sensations with that word. However, the man always uses the word in a way that fits our normal use insofar as it accompanies the usual symptoms of pain. He says he is in pain after stubbing his toe and says that the pain is gone after taking painkillers etcetera. Wittgenstein says in response to such a case that a cog that can be turned without moving anything else is not a part of the mechanism.

Nevertheless, Wittgenstein is happy to admit that 'pain' expresses a sensation. In section 244 he in fact offers an account of how words refer to sensations. He says that words may refer to sensations by being connected with the primitive and natural expressions of those sensations. If 'pain' refers to a sensation, it is because a sensation gives rise to the natural expressions of pain. But if there were no particular sensation that fulfilled that role "I am in pain" would not refer to a sensation. In this way "I am in pain" does not simply express a relation between objects.

The case with a first-personal fact is different to the case of pain thus understood. A first-personal fact is a reason for the behaviour it brings about. If there were no pain sensation and no first-personal fact, then we must conclude that the behaviour was not a response to these objects. I did not stop because I was making a mess and I did not scream because I was in pain. But, because the first-personal fact was the reason for the behaviour it is associated with this would also mean that that behaviour was unreasonable. If Wittgenstein's imagined scenario is right, the pain behaviour would not be rendered unreasonable by the lack of the pain sensation. Conversely, if we think that the pain behaviour would be rendered unreasonable we ought to think that the sensation is not the idle wheel one might imagine.

The PLA as commonly understood does not undermine my reference to first-personal facts because its argument is unreliable. PLAI survives these criticisms, but it does not undermine first-personal facts which are not private in the relevant strict sense. First-personal facts are in a manner knowable generally. They are evinced in things besides themselves and even if one's first-personal facts are not known to others as the content of their propositional attitudes they are known in their practices.

10.5 Summary

In this chapter I have briefly given support to the argument from action by making it clear that it supports conclusions which have independent plausibility, namely the tensed theory of time and emergentism about the self. I have also defended the argument from criticisms that try to find fault with the conclusions of the argument.

The tensed theory is often criticized because of McTaggart's paradox. However, I have shown that McTaggart's paradox admits of various solutions compatible with the tensed theory. Crucially, the tense characteristics may be incompatible with one another but there is no reason to suppose any event possess more than one of these characteristics.

The idea that the facts captured by someone's first-personal beliefs/utterances cannot be fully grasped by anyone else might appear to conflict with Wittgenstein's argument that a private language is impossible. However, I have shown that a number of interpretations of Wittgenstein's argument are inadequate. Moreover, first-personal facts are not private in the strict sense Wittgenstein's argument is concerned with.

Notes

1. Cf. Shoemaker (1969).
2. Cf. e.g. Broad (1921) (though he later changed his mind adopting a tensed theory, e.g. Broad (1923)), Blake (1925), Braithwaite (1928), Mellor (1998), Le Poidevin (1991), and Oaklander (2004).
3. Russell's definition refers to the truth-values of propositions, however, one needn't go into these complications (e.g. Le Poidevin (1998, 16)).
4. Cf. e.g. Le Poidevin (1991) and Bourne (2006). Lowe (1987a) offers a rather different asymmetry: one's spatiotemporal route must follow a particular temporal order, but not necessarily a particular spatial one. Mellor (1981) instead notes that objects are extended through space by having spatial parts, but they are not extended through time by having temporal parts.
5. For a rather different, arguably tensed, account of change see Lowe (2006a).
6. Some deny A2 on the grounds that only the present exists, e.g. Bourne (2006). But, I don't think a denial of A2 entails presentism.
7. This accommodates both the token-reflexive and date accounts of the truth-conditions because in order to be simultaneous with an utterance *e* must occur at the time of the utterance.
8. For examples of those who have adopted such harmless infinites or circles see Smith (1993, 1994) and Schlesinger (1994a, 1994b).
9. Lowe (1987a). For discussion of this point see also Lowe (1987b, 1992, 1993), MacBeath (1988), Le Poidevin and Mellor (1987), and Le Poidevin (1991, ch. 2, 1993).
10. This would mean that we cannot describe change as McTaggart does, but I have argued that this is no obstacle.

11. Cf. Wilson (2015).
12. I say more about what form this notion of dependence should be taken to have in (Pearson Forthcoming).
13. I will speak in this manner but I do not wish to enter disputes as to whether or not Wittgenstein actually argued the case or was merely trying to get us to see right etc. (cf. Canfield (2001)).
14. Kripke (1982) is an exception.
15. I do not intend my following distinctions to be exhaustive or exclusive of one another. Further, the references I give are not intended to pigeonhole writers whose views are often far more complicated than my distinctions might suggest.
16. Cf. Cockburn (2001, ch. 4). Though Cockburn makes this point in order to show how the case of 'S' differs from our normal use of sensation words such as 'pain'.
17. Cf. Canfield (2001).
18. The former of these views, U3, is perhaps held by those Kenny dismisses. The latter view, U4, is perhaps held by Kenny himself (Kenny (1973, ch. 10) especially pp. 191–192) and in a rather different form by Canfield (2001) insofar as Canfield questions the possibility of continuing a practice.
19. Cf. Pears (1988, ch. 13) and Glock (1996, entry 'The Private Language Argument').
20. Cf. Hacker (1987, 109).
21. Cf. Malcolm (1986).
22. Cf. Canfield (1991).
23. Cf. McGinn (1997, ch. 4) and Candlish (1980).
24. Descartes (1641/1974) is a prime example of this way of thinking.
25. E.g., this idea is criticized by Alston (1971) and Gertler (2003).
26. I am following Law (2004) in this paragraph and the three below it.
27. Blackburn (1984), see also Bain (2004).
28. A number of other and sometimes more explicit quotes from Wittgenstein's manuscripts can be found in Canfield (1996).
29. Cf. e.g. Wittgenstein (1953/2001, pt. I, sect. 257).
30. Cf. e.g. Wittgenstein (1953/2001, pt. I, sect. 260 and 268).
31. Cf. e.g. Wittgenstein (1953/2001, pt. I, sect. 270, 271, and 293).
32. A first-personal fact is a fact only captured by first-personal beliefs or utterances.

References

Alston, W. (1971). Varieties of Privileged Access. *American Philosophical Quarterly, 8*(3), 223–241.

Bain, D. (2004). Private Languages and Private Theorists. *The Philosophical Quarterly, 54*, 427–434.

Blackburn, S. W. (1984). The Individual Strikes Back. *Synthese, 58*, 281–302.

Blake, R. M. (1925). On Mr Broad's Theory of Time. *Mind, 34*, 418–435.

Bourne, C. (2006). *A Future for Presentism*. Oxford: Oxford University Press.

Braithwaite, R. B. (1928). Time and Change. *Proceedings of the Aristotelian Society, Supplementary Volume, 8*, 162–174.

Broad, C. D. (1921). Time. In J. Hastings (Ed.), *The Encyclopedia of Religion and Ethics* (pp. 334–345). Edinburgh: T & T Clark.

Broad, C. D. (1923). *Scientific Thought*. London: Kegan Paul, Trench, Trubner & Co.

Candlish, S. (1980). The Real Private Language Argument. *Philosophy, 55*, 85–94.

Canfield, J. V. (1991). Private Language: Philosophical Investigations Section 258 and Environs. In R. L. Arrington & H.-J. Glock (Eds.), *Wittgenstein's Philosophical Investigations* (pp. 120–137). London: Routledge.

Canfield, J. V. (1996). The Community View. *Philosophical Review, 105*, 469–488.

Canfield, J. V. (2001). Private Language: The Diary Case. *Australasian Journal of Philosophy, 79*, 377–394.

Cockburn, D. (2001). *An Introduction to the Philosophy of Mind*. Basingstoke: Palgrave.

Descartes, R. (1641/1974). Meditations on the First Philosophy. In J. Veitch (Trans.), *The Rationalists* (pp. 99–174). New York: Anchor Books.

Gertler, B. (2003). Introduction. In B. Gertler (Ed.), *Privileged Access* (pp. xi–xxii). London: Ashgate.

Glock, H.-J. (1996). *A Wittgenstein Dictionary*. Oxford: Blackwell Reference.

Hacker, P. M. S. (1987). *Appearance and Reality*. Oxford: Blackwell.

Hanfling, O. (1984). What Does the Private Language Argument Prove? *The Philosophical Quarterly, 34*, 468–481.

Kenny, A. J. P. (1973). *Wittgenstein*. London: Allen Lane The Penguin Press.

Kripke, S. (1982). *Wittgenstein on Rules and Private Language: An Elementary Exposition*. Cambridge, MA: Harvard University Press.

Law, S. (2004). Five Private Language Arguments. *International Journal of Philosophical Studies, 12*, 159–176.
Le Poidevin, R. (1991). *Change, Cause, and Contradiction: A Defence of the Tenseless Theory of Time*. New York: St. Martin's Press.
Le Poidevin, R. (1993). Lowe on McTaggart. *Mind, 102*, 163–170.
Le Poidevin, R. (1998). The Past, Present, and Future of the Debate about Tense. In R. Le Poidevin (Ed.), *Questions of Time and Tense* (pp. 13–42). Oxford: Oxford University Press.
Le Poidevin, R., & Mellor, D. H. (1987). Time, Change, and the 'Indexical Fallacy'. *Mind, 96*, 534–538.
Lowe, E. J. (1987a). The Indexical Fallacy in McTaggart's Proof of the Unreality of Time. *Mind, 96*, 62–70.
Lowe, E. J. (1987b). Reply to Le Poidevin and Mellor. *Mind, 96*, 539–542.
Lowe, E. J. (1992). McTaggart's Paradox Revisited. *Mind, 101*, 323–326.
Lowe, E. J. (1993). Comment on Le Poidevin. *Mind, 102*, 171–173.
Lowe, E. J. (2006a). How Real is Substantial Change? *The Monist, 89*, 275–293.
Lowe, E. J. (2006b). Non-Cartesian Dualism and the Problem of Mental Causation. *Erkenntnis, 65*, 5–23.
Lowe, E. J. (2008). *Personal Agency*. Oxford: Oxford University Press.
MacBeath, M. (1988). Dummett's Second-Order Indexicals. *Mind, 97*, 113–116.
Malcolm, N. (1986). *Nothing is Hidden*. Oxford: Blackwell.
McGinn, M. (1997). *Wittgenstein and the Philosophical Investigations*. London: Routledge.
McTaggart, J. M. E. (1908). The Unreality of Time. *Mind, 68*, 457–474.
McTaggart, J. M. E. (1927). *The Nature of Existence* (Vol. II). Cambridge: Cambridge University Press.
Mellor, D. H. (1981). *Real Time*. Cambridge: Cambridge University Press.
Mellor, D. H. (1998). *Real Time II*. London: Routledge.
Oaklander, N. (2004). *The Ontology of Time*. Amherst: Prometheus Books.
Pears, D. F. (1988). *The False Prison* (Vol. 2). Oxford: Oxford University Press.
Pearson, O. (Forthcoming). Emergence, Dependence, and Fundamentality. *Erkenntnis*. https://doi.org/10.1007/s10670-017-9895-1
Prior, A. N. (1967). *Past, Present and Future*. Oxford: Oxford University Press.
Russell, B. (1903/1996). *The Principles of Mathematics*. London: W. W. Norton.
Schlesinger, G. (1994a). How to Navigate the River of Time. In N. Oaklander & Q. Smith (Eds.), *The New Theory of Time* (pp. 229–231). London: Yale University Press.

Schlesinger, G. (1994b). The Stream of Time. In N. Oaklander & Q. Smith (Eds.), *The New Theory of Time* (pp. 257–285). London: Yale University Press.

Shoemaker, S. (1969). Time Without Change. *Journal of Philosophy, 66*, 363–381.

Smith, Q. (1993). *Language and Time*. Oxford: Oxford University Press.

Smith, Q. (1994). The Infinite Regress of Temporal Attributions. In N. Oaklander & Q. Smith (Eds.), *The New Theory of Time* (pp. 180–194). London: Yale University Press.

Wilson, J. (2015). Metaphysical Emergence: Weak or Strong? In T. Bigaj & C. Wüthrich (Eds.), *Metaphysics in Contemporary Physics* (pp. 251–306). Amsterdam and New York: Rodopi.

Wittgenstein, L. (1953/2001). *Philosophical Investigations* (G. E. M. Anscombe, Trans.). Oxford: Blackwell.

11

Closing Remarks

In the preceding chapters I have provided arguments to show that the tensed theory of time and emergentism about the self are correct. These arguments grew out of a consideration of Prior's thank goodness argument and Perry' argument for the essential indexical. In this chapter I will show how my treatment of Perry's argument differs from many of those found in the literature. I will argue that although I draw conclusions from this argument which are not commonly drawn this is a result of my emphasizing an aspect of Perry's argument not often emphasized.

11.1 The Problem of the Essential Indexical

In presenting his argument Perry (1979) refers to the *problem* of the essential indexical and I too have treated his argument as providing a problem. However, the problem that I consider is different from that which Perry and a number of subsequent writers have addressed. Perry raises his argument as a problem for what he calls the *traditional* view that belief is a relation to a proposition. This is the view that when I believe that Perry is making a mess I am in the believing relation to the proposition that *Perry is making a mess*.

Olley (F.O.C.H.) Pearson, *Rationality, Time, and Self*,
https://doi.org/10.1007/978-3-319-71973-3_11

According to the traditional view[1] propositions have a particular nature: they have a determinate and unchanging truth-value. Further, two propositions are the same only if they share the same truth-value, they attribute the same properties or relations to the same objects, and they involve the same concepts. The propositions that *grass is green* and that *snow is white* meet the first condition but fail the second two. The propositions that *the home of the rules of golf is in Scotland* and that *St Andrews is in Scotland* meet the first two conditions but fail the third.

The question Perry raises is the following:

> Q1, what proposition does an agent believe when that agent has an indexical belief?

The argument for the essential indexical makes this problematic because we know that in certain cases an agent can have all the non-indexical beliefs that one could have and still fail to have a requisite indexical belief. This implies that the agent can have a belief relation to all of the propositions captured by non-indexical utterances and not have one to a proposition expressed by an indexical utterance. For example, I can believe 'Olley is making a mess' and 'the only son of Felicity is making a mess' but not know 'I am making a mess' because I can fail to know 'I am Olley' and 'I am the only son of Felicity'. Furthermore, when I believe 'I am making a mess' I believe something quite different to that which is believed by anyone else when he or she believes 'I am making a mess'.

This problem is not easily resolved by altering our notion of proposition. For example, we might consider relativized propositions which are only true at a particular index[2] such as the proposition that *I am making a mess*. This is a proposition only true at me when I am making a mess and only true at someone else when he or she is making a mess. In addition to being believed such relativized propositions must be believed true at an index. I can believe the proposition that *I am making a mess* is true of Perry but not of me if I believe that Perry is making a mess but I am not. But, it follows from this that I must believe that the proposition that *I am making a mess* is true at me when I believe I am making a mess. Further, because I can fail to know that I am Olley etcetera, I must believe 'the proposition that I am making a mess is true at *me*' or some other belief that picks me out with a first-personal indexical and not any non-first-personal referring expression. Therefore, the essential indexical reappears and our problem is not removed.

The problem is also not removed if we turn from talking of what might be termed *de dicto* beliefs to talking of *de re* beliefs. In *de re* beliefs an object stands in a special relation to the believer which enables the believer to refer to it without necessarily picking out anything particular about it. For example, when I see a person and believe of that person that she is tall without my having to note any individuating quality of the person.[3] This move to *de re* beliefs does not solve the problem because one is able to have a *de re* belief of oneself that one is making a mess without believing 'I am making a mess'. As Perry notes, I could see myself in a mirror making a mess and believe of the person I see that he is making a mess but fail to know that I am the person that I see.

Perry proposes to solve the problem by differentiating the proposition believed from the belief state. This means that two beliefs can be different despite picking out the same proposition because they nonetheless manifest different belief states. The answer to Q1 is hence that an indexical belief will pick out the same proposition as a related non-indexical belief. When I believe 'I am making a mess' and 'Olley is making a mess' my beliefs concern the same proposition, namely the proposition that *Olley is making a mess*. However, the beliefs manifest different belief states. When Perry believes 'I am making a mess' he will be in a belief state of the same type as I am in when I believe 'I am making a mess'. However, Perry's belief will pick out a different proposition from mine, namely the proposition that *Perry is making a mess*. Moreover, the actions of an individual are responsive to the belief states the individual is in rather than the propositions the individual believes. Therefore, I act differently when I have an indexical belief than when I have a non-indexical belief that picks out the same proposition.

Perry's reference to belief states therefore plays two roles. It shows how two beliefs that pick out the same proposition can have a different cognitive significance and it shows how they can lead to different actions. Belief states thus answer a further question:

Q2, how can indexical and non-indexical beliefs pick out the same facts and yet have different cognitive significance and lead to different actions?

My question, however, differs from either Q1 or Q2, it is the following:

Q3, how can an action be rational when it involves an indexical belief though not when it involves any non-indexical belief?

I am led to Q3 by noting that Perry's argument for the essential indexical does not show that indexical beliefs are essential for a particular bodily movement or doing of an agent. Rather, indexical beliefs are specifically essential for a particular rational action of the agent. One can answer Q1 and Q2 without answering Q3 because people can act irrationally. Moreover, one can give a story to meet Q1 and Q2 that pays no attention at all to the rationality of an action.

Perry's reference to belief states, even if it could answer Q1 and Q2, is no answer to Q3. We have seen that rationality demands that a response to Q3 must involve recognizing that the indexical and non-indexical beliefs pick out different facts and hence different propositions.[4] This is not surprising given that I believe Perry was not concerned with Q3. Nevertheless, Q3 derives from the argument for the essential indexical and therefore Perry has failed to address this important element of the argument. Conversely, a proper answer to Q3 does address Q1 and Q2. If indexical beliefs pick out different facts from non-indexical ones then the two will likely involve different propositions and inevitably have a different cognitive significance therefore leading to different actions. We needn't answer Q2 directly rather we dissolve the need to answer it by accounting for the data differently.[5]

11.2 Other Responses to Perry's Problem

Many[6] of those who have considered the argument for the essential indexical have focused, as Perry did, on Q1 and Q2 or the concerns they reflect. These concerns are also reflected in many accounts of the semantics of indexicals. I find this unsurprising given the origins of the argument in a consideration of propositions and the fact that the questions are very interesting in their own right. I will now briefly outline some of the other major responses that have been given and will indicate how they meet Q1 and Q2 but fail to meet Q3. I will not attempt to consider all accounts that have been given of indexicals. I merely wish to make plausible the idea that discussions often fail to consider Q3. This will provide an account of why my conclusions differ from many other responses to the argument for the essential indexical. Perhaps my conclusions have been neglected because Q3 has been neglected.

Lewis[7] answers Q1 by denying that beliefs are relations to propositions at all. Rather, when an agent has a belief that agent attributes a property to himself or herself. This enables Lewis to account, in part, for the argument for the essential indexical because he takes properties to be more fine-grained than propositions. For Lewis propositions are sets of possible worlds. For example, the proposition that *Perry is making a mess* will be the set of all the worlds in which Perry is making a mess. The proposition that *Olley is happy* will be the set of worlds in which Olley is happy and so on. Properties, however, cut across worlds as well as between them. The property of *redness*, for example, is the set of all those entities in all possible worlds which are red. Lewis has an abundant view of properties so that, for example, *inhabiting a world in which Perry is making a mess* is a property. Because of this there will be a property for each proposition and there will be many more properties besides.

According to Lewis, when I believe 'Durham is in England' I self-ascribe the property of *living in a world in which Durham is in England.* I narrow down the world I live in to being a world from this set. Similarly, when I believe 'Olley is making a mess' I self-ascribe the property of *living in a world in which Olley*[8] *is making a mess.* These are beliefs that could easily be accounted for with reference to propositions. On the other hand, when I—Olley—believe 'I am making a mess' I self-ascribe the property of *making a mess.* This property cannot be captured by a proposition (hence the problem for the traditional view). This difference in property self-ascribed accounts for why my first-personal belief and my third personal belief have a different cognitive significance and lead to different actions answering Q2.

Nevertheless, Lewis must say something more to account for the role of tensed beliefs. He must say more to account for how my 2pm belief 'it is 2pm now' differs from any tenseless belief of mine because the tensed and tenseless beliefs have different cognitive significance and play different roles. For example, my 2pm belief of 'it is 2pm now' must differ from my belief '2pm is 2pm'. The two beliefs cannot both simply be the self-ascription of the property of *living in a world where 2pm is 2pm* because they play different roles. Lewis accounts for this by saying that people are not continuants but are in fact composed by time-slices each of which exists at only one time.[9] When I know 'it is 2pm now' my 2pm time-slice self-ascribes the property of *existing at 2pm*. This is quite different from it self-ascribing the property of *living in a world where 2pm is 2pm*. With these tools one can also account for the difference between my believing whilst in London that 'London is here' and my believing 'London is London'. In the former I self-ascribe the property of *existing in London* and in the latter I self-ascribe the property of *existing in a world where London is London*.

In this way Lewis can answer Q1 and Q2. Indexical beliefs will involve a relation to a property—not a proposition—and they will differ from non-indexical beliefs by differing in respect to these properties. Nevertheless, Lewis's account does not provide an adequate answer to Q3. On Lewis's account the set of entities that one is related to in a particular first-personal belief might be different to the set of entities one is related to in a non-first-personal belief but these sets of entities are not reasons to act. Reasons to act are states of the actual world. Reasons are facts not sets of possible entities. We know from Chaps. 7 and 9 that in order to answer Q3 Lewis must say that an indexical belief captures facts not captured by any non-indexical belief. But this is not a result of Lewis's position (thus Lewis's answer to Q2 is that the beliefs involve different properties). I am a member of the set of entities that have the property of *making a mess* in virtue of the fact that I actually make a mess. My actually making a mess is also part of that in virtue of which I am a member of the set of entities that inhabit a world in which Olley is making a mess. There is a reason that the sets converge in the actual world. The same fact underwrites both sets and property instantiations. Moreover, all of the facts that occur in a possible world could be fully picked out and known non-first-personally. Therefore, first-personal beliefs do not capture facts about a world that non-first-personal beliefs miss on Lewis's view.

Chisholm[10] offers an account very similar to Lewis's except Chisholm does not posit time-slices of people. He instead recognizes that tensed beliefs capture something about the world not captured by tenseless beliefs. In this way he can go part way towards an answer to Q3 (though he does not consider the role of rationality as I have). However, his account remains insufficient as he does not recognize that first- and non-first-personal beliefs must capture different facts.

Stalnaker (1981) responds to Perry's argument for the essential indexical by saying that we can recognize there to be more propositions relevant to a belief than Perry appreciates and with which we can characterize the beliefs at issue. The propositions Perry recognizes can be thought of as functions from a possible world to a truth-value—*horizontal* propositions. For example, the proposition that *Perry is making a mess* is a function that gives the value true for any argument which is a world in which Perry is making a mess and the value false for any other argument. However, Stalnaker suggests that we can recognize further propositions which can be thought of as functions from pairs of possible worlds to truth-values or as functions from possible worlds to horizontal propositions. These are *diagonal* propositions. For Stalnaker these diagonal propositions can characterize certain beliefs. For example, if I am willing to accept the utterance "Olley is Olley" but am unsure about the utterance "I am Olley" what I do not know is the diagonal proposition arising for the utterance "I am Olley". I do not know the proposition that is true in a world if and only if the proposition expressed by the utterance "I am Olley" in that world is true in that world. This is different from the proposition expressed by "I am Olley" in this or any other world. This can be seen because the proposition expressed by "I am Olley" in this world—the proposition *Olley is Olley*—is true in all possible worlds. But if that utterance had been made by me in a world in which 'Olley' referred to someone else like Perry, it would be the proposition that *Olley is Perry* which is false in all possible worlds. The diagonal proposition, on the other hand, would be true in the actual world but false in the possible world in which Perry is known by the name 'Olley'. However, Stalnaker stresses that these diagonal propositions needn't be thought of as a special variety of propositions because they too could be the horizontal propositions of different utterances.

Stalnaker thus tries to answer Q1 and Q2 by referring to a further set of propositions which are associated with the beliefs at issue.[11] However, this is no answer to Q3 because, as with Lewis above, someone possessing all possible non-indexical knowledge would thereby know all of the facts—the states of our world, the reasons—that he or she could in an instance of indexical knowledge. In fact Stalnaker (1999) appears to agree as much at a later date when he emphasizes that what is crucial about an indexical belief is not so much what one knows in that belief but one's perspective upon that content.

These failings are not surprising as Perry, Lewis, Chisholm, and Stalnaker did not try to address Q3. They did not consider the nature of rationality. But with Q3 in mind they are quite obvious. One's brain states or the properties or propositions one is related to in a belief are not relevant to the rationality of an action arising from those beliefs insofar as they can vary independently of the facts that one is aware of in those beliefs. The rationality of one's action is concerned with its being an agential response to the world. It does not matter how one is composed or how one's response to the world is mediated.

Evans (1982) meets the concerns of Q1 and Q2 in part through recognizing there to be more propositions than Perry was willing to accept. For example, in my first-personal beliefs I will be related to propositions that no one else can be related to.[12] To this extent the comments I just made in regard to the other writers apply equally here. But Evans also emphasizes (perhaps more so than the others) the relations between beliefs and information and dispositions to act. For Evans these relations are in part constitutive of beliefs. For example, my first-personal and third-personal beliefs just are differently responsive to sources of information and lead to different actions.

This invites the idea that, for example, my belief 'I am making a mess' is essential for my action of stopping because a belief which leads to this action just is an 'I am making a mess' belief.[13] However, this cannot be correct because, as I have already argued, I could instead act on a third-personal belief if I act irrationally. The point I wish to make clear here and that I take to be a failing of Evans' is that the ties he makes between beliefs and actions are only plausible for rational actions. A belief cannot be defined in terms of the actions it leads to because it might lead to any action. Nevertheless a particular belief will fit only a specific set of rational actions. However, taking this association with rational actions to be partly constitutive of a belief does not account for why it and the rational actions fit together in this way. It does not answer Q3. At best it says that there is no explanation, "here our spade is turned". But, it is fair to think that there ought to be an explanation here. Moreover, I have already provided one. The beliefs and rational actions co-vary in this way because a belief is an awareness of particular facts and these facts are j-reasons for a particular set of rational actions.

11.3 Conclusion

My intention in this book has been to provide an argument for the tensed theory of time. I have done this by focussing on issues already familiar within the debate about metaphysical tense, namely the need for tensed beliefs in action and emotion. I showed that previous discussions of these issues were inadequate as they failed to account for the rationality of the actions and the appropriateness of the emotions. This is unsurprising given that rationality and appropriateness were often not considered in detail.

I examined the nature of rational action and showed that the rationality of an action stems solely from the reasons it is performed for and that the role of belief in an action is simply to provide the agent with an awareness of these reasons which are facts. This account of rational action fits the contemporary status of the field and is defended and plausible quite independently of the issues surrounding metaphysical tense. However, it does entail that tensed beliefs must capture facts that tenseless ones do not given that tensed beliefs are often required in rational actions.

In a parallel manner I defended a cognitive view of emotions according to which emotions are had for reasons. These reasons are the facts represented in the cognitive component of an emotion. This view of emotions brings in insight from the consideration of reasons and rationality. This view of emotions also entails that tensed emotions capture facts not captured by tenseless ones.

These arguments for the tensed theory of time have a general form. They entail that, for example, if two sentences which differ only in containing different co-referring names can play a different role in rational action (or appropriate emotion), then these sentences must capture different facts. These implications are not problematic and in fact have independent plausibility stemming from theories of reference and accounts of cognitive significance. For example, a descriptive theory of reference will associate different facts with different co-referring names and, more generally, it is tempting to think that two co-referring names will have a different cognitive significance precisely because the two have different facts associated with them.

The tensed theory of time has been criticized on the basis of McTaggart's paradox. However, I have shown that these criticisms are ill-founded because the tensed theory needn't lead to the paradox envisioned because the tensed theory does not entail that anything is at once past, present, and future. Besides metaphysical tense, my arguments support emergentism about the self by showing that a person's first-personal beliefs will capture facts not captured by any other beliefs. This conclusion appears to conflict with Wittgenstein's private language argument. However, again, this appearance is in-conclusive. Certain interpretations of Wittgenstein's argument present poor arguments. On the other hand, stronger interpretations show that these first-personal facts are not private in the manner that concerned Wittgenstein. They are knowable to other people. If they are not the referents of their propositional attitudes, they are nevertheless recognized by them in their practices.

Thus the primary value of this book is to provide a novel argument to establish a disputed view in the philosophy of time, namely the tensed theory of time. This conclusion is reached in a manner that leaves it open exactly how this tensed aspect of time ought to be understood. For example, it is left open whether there are tensed properties or a tensed aspect of existence. I take this to be an advantage of my argument which affords it a certain freedom. On the way to this conclusion I also provide an original tying together and embellishment of a number of arguments within the philosophy of action. Further, I use my understanding of these arguments to present a view of emotions which provides an insight into the inadequacies of some existing accounts of emotions. These positions also provide a novel argument to establish emergentism about the self.

If one were to disagree with my accounts of rational action and appropriate emotion, then he or she might in this way try to preserve a denial of the tensed theory of time and emergentism about the self. However, such an individual ought still to appreciate the link that I have shown to hold between certain intuitive accounts of rationality and appropriateness, and the tensed theory of time and emergentism about the self. I have shown that one cannot decree time or self to have a particular nature independently of considering the nature of other facets of our lives, and specifically normative ones. Similarly, it follows from what I have argued that one cannot decree the meaning of indexicals to have a particular nature independently of considering the nature of these normative issues.

Notes

1. Perry takes this view of propositions to be held by Frege though Evans (1981) thinks that Perry is sometimes unfair in his treatment of Frege's position (cf. Perry (1977)).
2. The indexes at issue here are specifically ones within a world, e.g., a context consisting of an agent, a time, and a place. It is quite common to think that propositions only have a truth-value relative to possible worlds. E.g., the proposition that *Olley is making a mess* will be true in all those worlds where I make a mess but not in any others.
3. There are a number of different accounts of *de re* beliefs some of which would involve one needing to know something particular about the referent but most agree that it is one's special relation to the referent rather than the knowing of the quality which distinguishes the *de re* from the *de dicto* (cf. Kaplan (1968)).
4. If propositions are traditionally understood this follows. Though I do not want to be committed to this idea or interpretation of propositions. Perhaps a single I-proposition could pick out different facts for different people.
5. Perry expands upon this view in more recent work (e.g. Perry (1997, 1998, 2006) in doing so he makes more propositions and facts relevant

to the indexical beliefs (propositions concerning meaning and context). But it is clear that unless these are propositions that give facts non-indexical beliefs cannot capture, which they do not, then they will not be adequate to answer Q3.

6. There are of course exceptions. Perhaps most notable is Castaneda, whose work inspired Perry and who goes on to give a more detailed account of action (cf. Castaneda (1990, 1992)). Parfit (1987) is another counterexample. However, I have addressed these accounts by defending a particular view of rational action in Chaps. 4, 5, and 6.
7. Lewis (1979). See Lewis (2001) for details regarding possible worlds, propositions, and properties.
8. Lewis believes that individuals exist in only one possible world but that they have counterparts that are relevantly similar and that exist in different possible worlds. To this extent my presentation is misleading. It might be better to speak of self-ascribing the property of living in a world in which someone called 'Olley' is making a mess. But this does not affect my argument, so for ease of presentation I write as I do in the main text.
9. Again I slightly misrepresent Lewis here because he believes that time slices can have a duration as long as they do not change in that time. However, again, this doesn't alter my argument so for simplicity I speak as I do. (Cf. Lewis (1983a, 1983b, 2001).)
10. Chisholm (1981). I am inclined to agree with Chisholm that people do not exist through time by being made up of time slices. For arguments to this effect see Lowe (1987, 1998, chs. 4 and 5).
11. I am not sure that this really works because I do not think it is clear that a diagonal proposition associated with a required indexical belief could not be the content of an appropriate non-indexical belief.
12. Which is not to say it is subjective (cf. Evans (1981)).
13. I do not mean Evans would say something this simplistic.

References

Castaneda, H.-N. (1990). Practical Thinking, Reasons for Doing, and Intentional Action: The Thinking of Doing and the Doing of Thinking. *Philosophical Perspectives, 4*, 273–308.

Castaneda, H.-N. (1992). Indexical Reference and Bodily Causal Diagrams in Intentional Action. *Studia Logica, 51*, 439–462.

Chisholm, R. (1981). *The First Person: An Essay on Reference and Intentionality*. Brighton: Harvester.

Evans, G. (1981). Understanding Demonstratives. In H. Parret & J. Bouveresse (Eds.), *Meaning and Understanding* (pp. 280–303). Berlin: de Gruyter.

Evans, G. (1982). In J. McDowell (Ed.), *The Varieties of Reference*. Oxford: Oxford University Press.

Kaplan, D. (1968). Quantifying In. *Synthese, 19*, 178–214.

Lewis, D. K. (1979). Attitudes *de Dicto* and *de Se*. *Philosophical Review, 88*, 513–543.

Lewis, D. K. (1983a). Survival and Identity. In *Philosophical Papers* (Vol. I, pp. 55–72). Oxford: Oxford University Press.

Lewis, D. K. (1983b). Postscript to 'Survival and Identity'. In *Philosophical Papers* (Vol. I, pp. 73–77). Oxford: Oxford University Press.

Lewis, D. K. (2001). *On the Plurality of Worlds*. Oxford: Blackwell.

Lowe, E. J. (1987). Lewis on Perdurance Versus Endurance. *Analysis, 47*, 152–154.

Lowe, E. J. (1998). *The Possibility of Metaphysics*. Oxford: Oxford University Press.

Parfit, D. (1987). *Reasons and Persons*. Oxford: Oxford University Press.

Perry, J. (1977). Frege on Demonstratives. *Philosophical Review, 86*, 474–497.

Perry, J. (1979). The Problem of the Essential Indexical. *Nous, 13*, 3–21.

Perry, J. (1997). Indexicals and Demonstratives. In B. Hale & C. Wright (Eds.), *A Companion to the Philosophy of Language* (pp. 586–612). Oxford: Blackwell.

Perry, J. (1998). Myself and I. In M. Stamm (Ed.), *Philosophie in Synthetisher Absicht (A festschrift for Dieter Heinrich)* (pp. 83–103). Stuttgart: Klett-Cotta.

Perry, J. (2006). Self. In D. M. Borchert (Ed.), *Encyclopedia of Philosophy* (Vol. 8, pp. 708–711). Detroit: Macmillan.

Stalnaker, R. C. (1981). Indexical Belief. *Synthese, 49*, 129–151.

Stalnaker, R. C. (1999). Introduction. In *Context and Content: Essays on Intentionality in Speech and Thought* (pp. 1–28). Oxford: Oxford University Press.

Index[1]

A

[1] Note: Page numbers followed by "n" refers to note.

Olley (F.O.C.H.) Pearson, *Rationality, Time, and Self*,
https://doi.org/10.1007/978-3-319-71973-3